A RABBIT IN THE CLOUDS

PSYCHOLOGY OF BELIEFS

Religions, Conspiracy Theories &

Pseudoscience

DANIEL CHABOT

Table of Contents

FOREWORD 3

INTRODUCTION 9

PART ONE THE MAKING OF BELIEFS 14

CHAPTER 1 THE MOTIVES AND PSYCHOLOGICAL FACTORS
....15

CHAPTER 2 THE SOCIAL MOTIVES AND FACTORS 28

CHAPTER 3 PERSONALITY TRAITS 42

PART TWO ADHERENCE AND COMMITMENT TO A BELIEF
SYSTEM 55

CHAPTER 4 THE ESCALATION OF COMMITMENT: FROM
HARMLESS BELIEF TO BELLIGERENT FUNDAMENTALISM ..56

CHAPTER 5 LIES, FAKE NEWS, RUMORS, AND PROPAGANDA
....70

CHAPTER 6 AN EXAMPLE OF BELIEF-MAKING 79

PART THREE ADDICTION TO BELIEFS AND THEIR POTENTIAL
DEVIATIONS 90

CHAPTER 7 THE OPIUM OF THE PEOPLE 91

CHAPTER 8 THE DEVOTED ACTOR 102

PART FOUR SCIENTIFIC THINKING AND CULTURE AS AN
ANTIDOTE TO IRRATIONAL BELIEFS 116

PREAMBLE 117

CHAPTER 9 THE MEANS OF ACQUIRING KNOWLEDGE 120

CHAPTER 10 DISTINGUISHING WHAT IS SCIENTIFIC FROM WHAT IS NOT 131

CHAPTER 11 SAFEGUARDS IN SCIENCE 140

CHAPTER 12 A FEW PRACTICAL TIPS FOR THE READER 149

CONCLUSION 157

APPENDIX 1 166

APPENDIX 2 DISTINGUISHING COINCIDENCE, CORRELATION, AND CAUSATION 169

APPENDIX 3 VALIDATING INFORMATION SOURCES 173

APPENDIX 4 CONSULTING PRIMARY SOURCES 176

APPENDIX 5 DO NOT HESITATE TO CONSULT FACT-CHECKERS 178

APPENDIX 6 ABUSE OF LANGUAGE..181

APPENDIX 7 RESISTING THE TRAP OF RELYING ON ANECDOTES 184

APPENDIX 8 BEING VIGILANT ABOUT OUR OWN COGNITIVE BIASES 188

APPENDIX 9 THE STEPS OF THE SCIENTIFIC METHOD 189

BIBLOGRAPHY 197

BY THE SAME AUTHOR

Popular Works in Psychology:

The Wisdom of Pleasure, Québecor Editions, 1st ed. 1991, 2nd ed. 2001.

Pleasure and Consciousness, Québecor Editions, 1st ed. 1993, 2nd ed. 2001.

The Magic of Pleasure, Québecor Editions, 1st ed. 1996, 2nd ed. 2002.

Cultivate Your Emotional Intelligence, Québecor Editions, 1998.

Translated into Portuguese (Pergaminho, 2000), Spanish (Mensajero, 2001), Italian (Punto d'Incontro, 2003), and Chinese (Energy, 2001).

Master Your Emotions Through Emotional Intelligence, Québec-Livres Editions, 2000.

Educational and Didactic Works:

Daniel Chabot and Michel Chabot, Emotional Pedagogy: To Feel in Order to Learn, Trafford Publishing, 2004.

Daniel Chabot and Michel Chabot, Pédagogie émotionnelle: ressentir pour apprendre (Emotional Pedagogy: Feeling in Order to Learn), Trafford Publishing, 2005.

Translated into Brazilian Portuguese (Sa Editora, 2008), Croatian (Educa, 2009), and Spanish (Alfaomega, 2009).

Daniel Chabot, Emotions and Adaptation, CCFD, 1996.

Methodology in the Human Sciences, Course Notes, Collège de Rosemont, 1998.

Psychology of Sport, Course Notes, Collège de Rosemont, 2022.

FOREWORD

Like many young people (and not-so-young ones), the great existential questions once appeared along the path of my life. And, like so many others, I searched for answers. With hindsight, I would say my journey followed two parallel yet distinct paths. One of them, which I would describe as experiential, led me to flirt with several spiritual and philosophical movements that were fashionable in the 1970s and 1980s. One of them captivated me more than the others, and I remained devoted to it for about forty years. The other, more rational path led me toward advanced university studies in psychology, which in turn brought me to a career teaching this subject at the college level (pre-university training in the social sciences).

So I explored the scientific literature on the psychological mechanisms involved in the creation and adherence to beliefs.

This book brings together the conclusions from those studies and proposes some answers to that question.

Important Clarifications

Although the central theme of this book deals with the psychology behind the formation and adherence to beliefs, I must emphasize from the outset that this does not exclude or oppose the fundamental right of each individual to believe in whatever they wish.

What is presented in these pages focuses not on the content of beliefs, but rather on the processes by which they are fabricated and accepted. And because beliefs directly influence our behaviors, I find it essential to take a rigorous, lucid look at the ingredients and mechanisms shaping them.

I must also stress that all the examples selected and presented in the following chapters are only meant to illustrate and highlight the concepts described by research on this subject. They do not necessarily reflect my personal views or opinions.

The Broad Reach of Beliefs

The word belief encompasses our opinions, convictions, values, principles, conceptions, and worldviews, among others.

Beyond the complexity and nuance involved in forming and adopting beliefs lies their crucial role, especially when they motivate and direct our behavior.

Over the past few decades, cognitive psychology has revealed the deep interaction between cognition and behavior. As Susan T. Fiske and Shelley E. Taylor note, "Cognitions give rise to behaviors... and behavior also gives rise to cognitions."

Therefore, we must move beyond the common claim that beliefs are "harmless" or purely a matter of personal choice. On the contrary, beliefs play a determining role, and their consequences extend far beyond the individual who holds them, they affect our relationships with others and the functioning of society as a whole.

History shows that beliefs and ideologies can foster social unity and cohesion. But it also shows that they can fracture that unity, divide societies, and even fuel extremes of fanaticism or violence.

A Rabbit in the Clouds

The contributions of the natural and health sciences to human progress are immeasurable and undeniable. Yet increasingly, the human and social sciences play a vital role in understanding human and social behavior, including the mechanisms underlying the creation of all forms of belief.

A Final Note Before Entering the Core Topic

You will notice, in the chapters that follow, that several sources refer to the COVID-19 pandemic. The reason is simple. Beyond the profound impact this pandemic had on minds worldwide, virtually every field of scientific research was touched by this historic, global event.

Among the many disciplines that took interest were those within the broad family of Human and Social Sciences (HSS): psychology, sociology, political science, history, anthropology, and more.

Typically, the study of social events, whether major or minor, such as the rise of religions or ideological movements, social revolutions, wars, conflicts, or epidemics, must rely on historical research based on indirect observations.

As François Simiand wrote as early as 1903:

"Because of their scope or nature, social phenomena can hardly be grasped through direct individual observation: hence the necessary recourse to indirect knowledge. Moreover, in social matters, the artificial experiment, produced at will under the scientist's eyes, is so rare and difficult to achieve that one must look for facts and cases of experience in the past of humanity: thus, the necessary use of the historical method."

However, what unfolded during the pandemic was directly observable in real time by researchers across disciplines.

They had access to massive data from laboratories and research centers around the world; they could design vast samples and study countless psychological, social, political, and economic issues.

Mass and social media played a more significant role than ever before, allowing information and beliefs of every kind to circulate at lightning speed, flooding, even submerging, the world. The handling of media and information thus became a subject of research.

Across the globe, researchers in the human and social sciences mobilized, and the number of publications exploded. Among them, seventy scholars from various fields joined forces to examine this immense, uncertain phenomenon. They published a first report in November 2020, followed by a revision in March 2021.

One of their key findings was:

"Very often, it is these uncertainties that feed the crisis, because they destabilize, overwhelm organizational response capacities, discredit authorities, and fuel major divergences among actors involved. Each crisis may have a cumulative effect on public trust. The current one illustrates this vividly, even radicalizes it. Beyond uncertainties, we face numerous, sometimes violent controversies, amplified by social and traditional media, which have provided continuous coverage saturated with numbers. In this movement, the boundary between 'fake news' and science, between legitimate and illegitimate sources of knowledge, tends to blur. Moreover, these controversies reflect an unprecedented politicization of public-health issues."

The creation and adoption of beliefs and theories, whether religious, political, superstitious, paranormal, pseudoscientific, or conspiratorial, rest largely on the same psychological mechanisms.

As Pascal Wagner-Egger, a researcher in social psychology at the University of Fribourg (Switzerland), notes, "Several studies in psychology have indeed found a statistical link between various paranormal, magical, or esoteric beliefs (astrology, telepathy, homeopathy, miracles, lucky charms, visits from extraterrestrials, etc.) and belief in conspiracy theories."

Many belief systems share similar characteristics, notably a conspiratorial mindset.

According to researchers from UNESCO's Chair in the Prevention of Radicalization and Violent Extremism:

"Conspiratorial thinking constitutes a worldview, an excessively suspicious way of seeing power or elites in general, and assumes that dark, malevolent forces are at work within social dynamics."

Beliefs and theories exist largely as forms of discourse, while conspiratorial thinking is both a predisposition to accept such discourse and an effect of that discourse, which can generate new suspicions among its audience, creating a feedback loop that draws adherents from doubt to conviction.

INTRODUCTION

The mental mechanism that makes us perceive the shape of a rabbit in the clouds is known as illusory pattern perception (IPP), defined as "the identification of a coherent and meaningful relationship among a set of random or unrelated stimuli."

According to certain psychologists, this perceptual bias is one of the key predictors of irrational beliefs, religious convictions, superstitions, and conspiracy theories.

But let us not move too fast. To understand this phenomenon better, let's take a look at a collective experience we all lived through in real time: the COVID-19 pandemic.

From its very beginning, we witnessed a flood of information of all kinds, spreading at breathtaking speed through traditional and social media. Many researchers in the human sciences were immediately struck by this wave of "fake news" crashing before their eyes, where falsehood blended with truth, where decontextualization distorted real facts, where half-truths fueled rumors, and where misunderstood or misinterpreted scientific knowledge was recycled to patch together a variety of speculative theories.

Yet such a phenomenon was neither new nor unique. History is full of examples in which the emergence of inexplicable events, especially catastrophes or crises that shake our lives and our emotions, drives human beings to seek explanations, to find meaning, or to assign causes to what happens.

And indeed, they do find them. The more disrupted our daily life becomes, the more likely strange, irrational, and implausible theories and beliefs are to appear.

Looking at the many societies, communities, and social groups that have populated the Earth, we find countless beliefs and theories offering explanations for the unknown, for life's mysteries, for sudden disasters, and even for mundane daily events. Billions of people adhere to them, gather around them, and build systems, institutions, and religions that withstand the test of time and repeated challenges.

This naturally raises the question: why is the human being so credulous?

The question extends far beyond today's popular conspiracy theories; it applies to all systems of belief, from small cults to vast religious, political, or ideological organizations.

And another question follows: beyond the need to believe, why this powerful need to convince others, to the point of engaging in ideological crusades, wars of religion, or even terrorist acts aimed at destroying those who do not share one's convictions and are blamed for humanity's misfortunes?

In a single lifetime, it is common to adopt and even defend various ideas, opinions, beliefs, and theories. Some beliefs are remarkably resilient and can withstand countless external challenges. However, this "protective system" is not completely airtight. Many people eventually recognize inconsistencies, contradictions, or gaps within their own belief systems, enough to shake their faith and push them toward deep self-examination.

Of Simplification and Complexity

Our biosphere, the whole of living organisms and their environments, is the ultimate achievement of staggering complexity. Every ecosystem and element of biodiversity interacts and depends on countless others.

The environmentalist James Lovelock proposed calling this whole the "symbiosphere" to highlight the interdependence among species and the unity they form.

It is as if we were all playing a vast, multidimensional game of chess.

In traditional chess, every time one player moves a piece, the opponent faces a finite set of possible responses but must choose only one. Each move triggers another. The game is sequential and linear, and the rules are clearly defined.

Reality, however, operates differently.

A single change, movement, or intervention can trigger a chain of reactions, instantaneous or gradual, simultaneous or alternating, controllable or inevitable. Systems can react, interact, or interfere cyclically, sequentially, linearly, or in cross-feedback patterns.

Sometimes, one intervention produces little or no effect because compensatory mechanisms step in to moderate, offset, or correct it.

A Rabbit in the Clouds

Human beings struggle not only to grasp complexity but also to account for it when thinking, deciding, or acting. We are limited in our ability to gauge the relevance, scope, and consequences of our choices and actions.

To cope, our minds rely on more economical cognitive mechanisms; we simplify complexity by:

- Taking mental shortcuts;

- Ignoring, eliminating, or subtracting numerous factors;

- Isolating those elements that seem most "obvious," familiar, or easy to handle;

- And once these are selected, drawing conclusions and generalizations that may be reasonable, or utterly irrational.

Let's be clear: as soon as we remove or ignore certain elements in order to simplify our understanding of a complex phenomenon, we must admit that our conclusions will necessarily be partial and offer only a distorted representation of reality.

Therefore, I invite the reader to keep in mind that our biosphere and universe are like an immense, multidimensional chessboard, an intricate web of interdependent, interconnected, and interactive factors.

Because of this overwhelming complexity (even anxiety-inducing complexity), countless psychological and sociological factors motivate and shape our interpretation of the world.

A Rabbit in the Clouds

Over-simplifying this complexity can create the illusion of understanding: we isolate a few obvious parameters, wrap them in heuristic hypotheses, and dress them in attractive theories or interpretations, then imagine we have grasped reality.

In other words, we open wide the gates to the fabrication and adoption of all kinds of beliefs.

*****

The purpose of this book is not to compile a ranking of beliefs, nor to judge their truth or value.

What motivated its writing stems from a deep desire to understand the foundations, motives, and consequences behind the formation of beliefs and the act of believing.

This search led me to a mountain of publications in the human sciences, particularly psychology, exploring what drives people to create, adhere to, and commit passionately to something that, in fact, may have little or no objective, verifiable evidence to support its authenticity.

No more, in truth, than there is any evidence of the actual existence of a rabbit in the clouds.

PART ONE
THE MAKING OF
BELIEFS

CHAPTER 1
THE MOTIVES AND PSYCHOLOGICAL FACTORS

We previously emphasized that it is impossible to know or understand everything about the complex world we live in. Yet, by its very nature, the human brain tolerates the unknown and uncertainty very poorly.

For this highly specialized organ, designed to process information, the absence of data is profoundly uncomfortable.

Think of our ancestors, who lived in hostile and dangerous environments. Accurately interpreting their surroundings was often a matter of survival. Despite lacking complete information, humanity adapted and endured.

Perception: The First Link in Knowing What Surrounds Us

In psychology, perception is defined as a sequential process of selecting, organizing, and interpreting the stimuli captured by our sensory receptors. Because we are constantly bombarded with an infinite number of stimuli, our brains must filter and select among them, determining which are relevant and which are not.

This selection depends on many physical, physiological, and psychological factors. From this, we understand that perception is inherently selective. Our brain never receives all available information. It must therefore organize and interpret partial data to reach the ultimate goal of perception: to give meaning to what we call "reality."

In short, we build our representation of reality from approximate, incomplete, and often insufficient information.

As a result, the brain must fill in these gaps to construct a coherent and secure picture of reality. Yet, because of its intrinsic limitations, we are condemned to function and adapt based on only a sketch of reality, while being flooded with a chaotic mass of stimuli.

The Intuitive Mode of Thinking

How, then, has humankind survived with such an arbitrary grasp of reality?

To answer that, let's look at how human thought operates.

According to many researchers, the human mind relies on two functional systems to process information about its physical and social environment.

This dual-process model distinguishes between:

- System 1, which is fast, intuitive, emotional, and heuristic-driven, and

- System 2, which is slow, analytical, rational, and deliberate.

For example, imagine someone who feels that a political leader is dishonest and corrupt. This impression, which emerges spontaneously, is a product of System 1, intuition, and emotional heuristics based on partial knowledge and vague interpretations.

A Rabbit in the Clouds

To verify or refute this belief, one must activate System 2, engaging in a deeper, rational analysis of factual data about that leader.

Both systems might reach the same conclusion (that the person is untrustworthy), but their paths of reasoning are radically different: one emotional and instinctive, the other analytical and evidence-based.

The intuitive system, therefore, operates quickly and effortlessly. When the brain must act or decide rapidly, it cannot account for every parameter. Speed comes at the expense of precision.

The faster we process information, the fewer stimuli we evaluate, and the greater the risk of error, bias, and distortion.

This does not mean intuition is useless.

As social psychologist Pascal Wagner-Egger (University of Fribourg, Switzerland) explains:

"... an intuitive, non-rational mode of thinking may have evolved as a survival mechanism for our species and its ancestors, who lived for hundreds of thousands, if not millions, of years as hunter-gatherers.

For instance, an intentionality bias, assuming danger behind any unexplained noise, could have saved lives. Even if most of these alarms were false (caused by the wind, harmless animals, or other humans), a few true ones, when the noise came from a predator or enemy, ensured survival. We still feel this bias today when walking in a dark forest or at night."

In short, by its very design, the intuitive mind is destined to make frequent mistakes.

As Wagner-Egger summarizes: "Our brain is gifted for survival, but far less so for truth."

Many studies show that beliefs of all kinds stem largely from this intuitive mode of thinking, combined with the simplification of analytical thought and the anxiety caused by uncertainty.

Once formed, such beliefs are then reinforced through rationalization: people selectively seek out evidence to confirm them.

Psychologist Jan-Willem van Prooijen and colleagues conclude that conspiracy theories emerge from System 1 (intuition) but are maintained and strengthened by System 2 (analysis) through cognitive biases that justify and defend them.

Illusory Pattern Perception

Our brains rely heavily on experience and memory to construct a functional representation of reality, a vital adaptive mechanism. But, as mentioned earlier, it is not free from error. In our hypercomplex world, it is impossible to perceive, decode, or understand everything.

A Rabbit in the Clouds

Figure. What do you see?

When we look at an abstract figure and recognize, among scattered spots and incomplete lines, the image of a soccer ball or a tiger's head, we are relying on prior knowledge of those shapes. Such harmless examples of pattern recognition are benign.

However, when we move beyond "rabbits in clouds" to the complex realities of life, with real consequences, problems arise. Our brains allow us to perceive what is real, but also, very often, to see things that do not exist.

In other words, our brains frequently deceive us.

This phenomenon extends far beyond visual perception; it influences our entire conception of the world. Because our minds struggle with ambiguity, mystery, and uncertainty, our interpretations of reality are constantly shaped by this intolerance for the unknown. This is one reason why humanity has produced such an abundance of beliefs, myths, legends, superstitions, and theories; they help us make sense of what we cannot directly perceive or explain. They reduce existential discomfort and anxiety caused by meaninglessness.

Research has shown that one of the main motives behind irrational and conspiratorial thinking is the need to understand and explain our complex world simply. Driven by this epistemic motivation, humans tend to connect random, unrelated events, seeing relationships where none exist.

In experiments by Jan-Willem van Prooijen and his team, participants prone to conspiracy beliefs were more likely to perceive patterns or structures in completely abstract images (such as works by Victor Vasarely or Jackson Pollock) and to infer connections between random data points.

In short, superstitions, myths, rumors, and beliefs often originate from this first cognitive link, the perception of illusory patterns. This phenomenon, also called apophenia, occurs when we assign meaning to meaningless things or perceive relationships among unrelated elements simply because of coincidence or chance.

Multiple studies have demonstrated a strong link between illusory pattern perception and belief in conspiracy, paranormal, or pseudoscientific ideas. When faced with ambiguity, humans create unfounded connections to fill gaps in understanding, sometimes giving birth to full-fledged belief systems.

In a series of experiments, Jennifer A. Whitson and Adam D. Galinsky found that participants who lacked a sense of control were significantly more likely to perceive illusory patterns, such as seeing images in visual noise, inventing correlations in stock data, imagining conspiracies, and developing superstitions.

The Intentionality Bias

The intentionality bias is the tendency to assume that events do not happen by chance, that there is always a deliberate, purposeful cause behind them.

Take the COVID-19 pandemic, for example.

At its onset, unanswered questions abounded:

Where did SARS-CoV-2 come from? What are the risks? How does it spread? Who is most vulnerable?

Faced with these uncertainties, many people adopted beliefs that malevolent human intentions were behind it: that the virus was engineered in Chinese laboratories; that it was a hoax serving pharmaceutical interests; or that it was part of a vast secret plot to reduce the world's population.

As the author notes:

"For decades, we have observed that human beings believe too quickly, based on a few cases, through an intuitive mode of thought inherited from our ancestors whose main goal was survival, and cling to their beliefs, individually and socially."

A Rabbit in the Clouds

The Need for Cognitive Closure

Another key epistemic motive is the Need for Cognitive Closure (NFC), the discomfort caused by ambiguity combined with a desire for a clear, definitive answer.

Psychologists Arie Kruglanski and Shira Fishman describe it as a "motivational stopping mechanism" that applies "brakes" to cognitive inquiry, leading to crystallized judgments. Because religious, mythic, conspiratorial, and paranormal systems provide simple answers to complex and uncomfortable questions, studies show that individuals with a high need for closure or a low tolerance for uncertainty are more likely to endorse intuitive, unverified beliefs.

In short, humans experience distress when faced with uncertainty or unresolved questions.

To relieve that discomfort, we often construct beliefs or theories that "fill the gap." The problem, however, arises when there is insufficient information to clarify or explain the phenomenon. The challenge becomes even greater when the event is complex, threatening, or stressful.

Under such anxiety, our intuitive system links unrelated events, human actions, or objects, producing cognitive structures designed to soothe discomfort. Thus arise the surges of myths, superstitions, and conspiracy theories that flood human culture.

A Rabbit in the Clouds

The Role of Emotions in Belief Formation

The unknown and the incomprehensible cause unease and a sense of lost control. These emotional reactions play a double role:

1. They intensify our urge to create beliefs or theories to ease discomfort, and
2. They reinforce our willingness to accept those beliefs for the comfort they provide.

As a result, humans often prefer emotional reassurance to objective accuracy, seeking comfort rather than truth.

The Attraction to Simplicity and the Sensational

Negative emotions tend to trigger chains of emotional and behavioral reactions, both individual and collective. Research shows that people are more captivated by the emotional thrill of discovering a mysterious or conspiratorial theory than by the intellectual labor of verifying it. It is far less exciting to engage in slow, rational analysis than to indulge in sensational, well-packaged stories. Psychologist Jan-Willem van Prooijen found that interesting, catchy information is easier for the brain to process than realistic but dull information.

This cognitive fluency, combined with emotional arousal, increases belief in sensational theories. As van Prooijen puts it, "The combination of mental ease and reduced rationality encourages belief that there may be truth in an entertaining conspiracy theory."

Further experiments by a team of European researchers showed three key findings:

1. Analytical thinking reduces belief in conspiracy theories.

2. Strong conspiracy belief correlates with lower analytical thinking.

3. Encouraging analytical thought experimentally decreases conspiratorial ideation.

Hence, promoting analytical reasoning is an effective way to counter widespread acceptance of conspiracy narratives.

Insights from Neuroscience

As mentioned earlier, the brain dislikes ambiguity and relies on inference to compensate for missing data. Inferences are deductions drawn from partial observations and personal knowledge or experience.

When those observations are uncertain or knowledge is limited, the brain's inferential system falters, often producing irrational or bizarre beliefs. A team of neuroscientists and psychiatrists from Hôpital Sainte-Anne and Université de Paris / Inserm investigated what happens when uncertainty is induced experimentally.

They used the drug ketamine, which blocks NMDA synaptic receptors, disrupting certain neural circuits and impairing cognitive function. Participants were asked to make decisions based on ambiguous visual information. Compared with placebo, ketamine produced both heightened uncertainty and premature decisions.

These hasty decisions might appear to stem from overconfidence, but researchers found the opposite: increased uncertainty prompted compensatory decision-making, committing prematurely in order to reduce ambiguity.

As neuroscientist Valentin Wyart noted:

"These premature decisions do not result from excessive confidence. On the contrary, they arise from elevated uncertainty, generating improbable ideas that reinforce themselves and resist contradiction."

This "compensatory control" mechanism mirrors what Aaron C. Kay and colleagues describe in their Compensatory Control Theory: when people feel a loss of personal control, they turn to external sources of order or certainty, religion, conspiracy, superstition, to regain a sense of stability.

*****

The following table summarizes the motives and psychological factors presented in this chapter. These are defined and illustrated using a common example:

PSYCHOLOGICAL MOTIVE	EXAMPLE
Intuitive Thinking Fast, emotion-based thinking guided by feelings and heuristics.	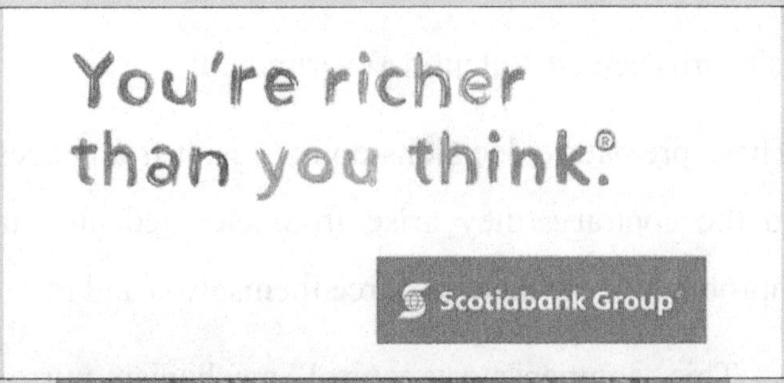 You're richer than you think.® Scotiabank Group
Illusory Pattern Perception (IPP) Assigning meaning or connections to random elements.	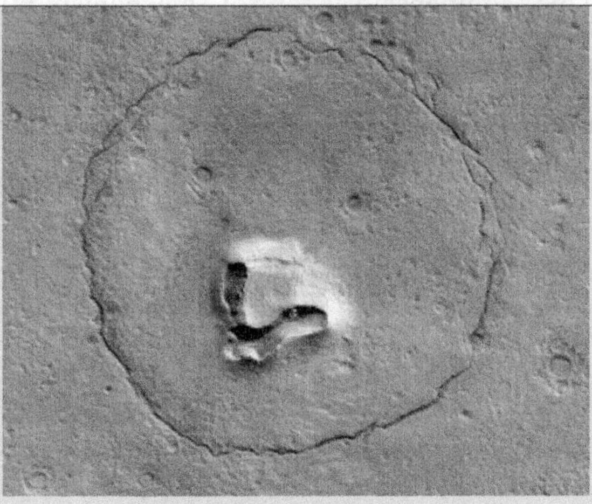 **A Bear Face on Mars.** Source : https://www.nasa.gov/image-article/a-bear-on-mars/
Intentionality Bias Assuming events result from deliberate intent (often malicious).	"Covid-19 is targeted to attack Caucasians and Black people" and "the people who are most immune are Ashkenazi Jews and Chinese…" Robert F. Kennedy Jr, jully 2023 Source : https://nypost.com/2023/07/15/rfk-jr-says-covid-was-ethnically-targeted-to-spare-jews/  NEW YORK POST RFK Jr. says COVID may have been 'ethnically targeted' to spare Jews

A Rabbit in the Clouds

Need for Cognitive Closure Discomfort with ambiguity and desire for definite answers.	« The Martian Face is compared to 'similar faces… constructed in civilizations on Earth. The faces are looking up at the sky because they are looking up to God.' Or the Face was constructed by the survivors of an interplanetary war that left the surface of Mars (and the Moon) pockmarked and ravaged. » Source : https://www.enterprisemission.org/mission.html	
Cognitive Inference Drawing conclusions from partial observations and experience.	« I know the human being and fish can coexist peacefully," he explained. » Bush was right, you know. History has no record of a major fish uprising. (The movie *Jaws* doesn't count.) Source : https://content.time.com/time/specials/packages/article/0,28804,1870938_1870943_1870961,00.html	*Top 10 Bushisms — Fish Are Friends*
Compensatory Control Seeking external systems to regain a sense of order or safety.	 Source : https://www.youtube.com/watch?v=irG1Qdae_g0	

CHAPTER 2
THE SOCIAL MOTIVES AND FACTORS

Between the motives and psychological factors discussed in the previous chapter and the social motives and factors we will examine in this one, there are certain connecting mechanisms, among them, the desire to have and maintain a positive view of oneself and of one's ingroup. It turns out that various beliefs and theories help many people achieve this goal.

The Need for Uniqueness

One of the great paradoxes of human nature is the constant oscillation between the need to resemble others and the need to be unique. On the one hand, humans are social beings who need to belong to a group of similar individuals (the ingroup); on the other hand, they feel the need to distinguish themselves from others (the need for uniqueness).

This need for uniqueness, which is closely linked to self-esteem, refers to the desire to stand out from others, to feel unique, to be nonconformist, and to experience a sense of independence from others.

As part of his doctoral research, Anthony Lantian, from the Université de Grenoble Alpes, studied the functional relationship between uniqueness and belief in conspiracy theories and observed that:

- The stronger a person's need for uniqueness, the more likely they are to believe in conspiracy theories.

- Those who believe in conspiracy theories are more inclined to think they possess rare and secret information about various social, political, or religious phenomena;

- A situational increase in the need for uniqueness tends to promote the formation of various beliefs.

Roland Imhoff and Pia Karoline Lamberty, from the University of Mainz in Germany, found that part of the motivation to embrace various irrational beliefs is linked to the need for uniqueness. These researchers conducted three studies that established a modest but consistent connection between the need for uniqueness and a general conspiratorial mindset, as well as the endorsement of specific conspiracy beliefs.

Imhoff and Lamberty emphasize that their findings confirm what other researchers had already observed, that individuals with a strong need for uniqueness tend to resist the majority and are more likely to yield to minority or marginal influences.

In one experiment, the researchers presented participants with a fictitious conspiracy. They found that this theory received more interest and support from participants with a strong conspiratorial mindset when it was described as being endorsed only by a small minority of people.

A Rabbit in the Clouds

According to the researchers, "these findings support the notion that conspiracy beliefs may be adopted as a way to achieve a sense of uniqueness."

Existential Motivations and the Theory of Compensatory Control

Existential motivations represent another social factor that plays an important role in the human quest for meaning amid life's many mysteries. They reflect a need for control and security, while also driving efforts to feel safe and at ease in an environment filled with ambiguity, uncertainty, and the unexplained.

The theory of compensatory control explains that individuals are fundamentally motivated to compensate for a loss of personal control by adopting a worldview that suggests the existence of some form of order behind the events that unfold around them. We can easily imagine that there are many strategies people use to regain the feeling of being in control of external events. From that point on, almost anything becomes possible.

Yannick Chatelain, Caroline Cuny, and Jamel Khenfer note that "while some of us possess the 'skills' and the capacity to cope with situations that threaten our sense of personal control, even when such situations are objectively uncontrollable, others do not have the same psychological resources. Some may seek an 'excessive' refuge in religion or in conspiracy theories."

According to the theory of compensatory control, people experience negative emotional states when they feel they lack control, and to reduce this unpleasant feeling, they tend to attribute greater power to external agents. Aaron Kay and his colleagues showed that when people's sense of personal control is threatened or weakened, they can compensate in several ways:

- By perceiving order in chaos (illusory pattern perception);

- By defending the legitimacy of sociopolitical institutions that provide control, or marginal groups that give the impression of control;

- By adopting beliefs in interventionist and superstitious powers (intentionality bias), such as mysterious, divine, or extraterrestrial forces, or in superstitions and conspiracy-type theories.

Kay and his co-authors also gathered evidence showing that compensatory control processes can help people reduce anxiety and discomfort caused by the feeling of having no control over events. In fact, it appears that it is the feeling of personal loss of control, rather than the general threat itself, that activates compensatory control mechanisms. They also found that different forms of compensatory control can substitute for one another; if one proves ineffective, another may work, or at least give the illusion of working.

We can easily imagine compensatory control at play in a wide range of circumstances. For example, many professional athletes adopt idiosyncratic rituals (entering the field with the right foot, avoiding the boundary lines, following compulsive routines, etc.); nearly all people and cultures around the world practice a wide array of collective customs and rituals, initiations, religious ceremonies, celebrations of birth, marriage, graduation, funerals, and more.

The feeling of lacking individual control often leads people to turn toward supernatural forces or external powers, such as religion, superstition, or conspiratorial beliefs.

Intentionality Bias

In the previous chapter, we defined what is meant by the intentionality bias, that is, the tendency to overestimate the role of deliberate and intentional motives on the part of someone (or some entity) when an event occurs. This tendency to explain events by attributing them to intentional causes becomes even stronger when the event, phenomenon, or behavior in question has negative consequences.

A good example of this is the number of theories that claimed to reveal the "real intentions" behind the COVID-19 pandemic:

- The virus was allegedly created in a laboratory in China and then deliberately released into the population.
- The pharmaceutical industry was supposedly behind the pandemic in order to sell vaccines and medicines.
- It was said to be a link between 5G and SARS-CoV-2.
- A secret New World Order allegedly sought to control the world.
- This same group supposedly aimed to reduce the global population.

- "Big Pharma" was said to be blocking inexpensive medicines and denying the beneficial or curative effects of certain natural products.

The Casual Attribution Bias

The causal attribution bias leads people to create explanations for events beyond their control, while rejecting the possibility that such events may be accidental, random, or simply unknown. These processes may also strengthen the connection between one's sense of personal control and adherence to various systems of belief or theories.

Indeed, numerous studies have shown that individuals who feel they have lost control, and/or have a strong need for control, are more likely to endorse conspiratorial or belief-based explanations, especially in the face of uncontrollable life events such as unemployment, financial hardship, natural disasters, wars, epidemics, and so on.

A research team from the Institute of Experimental Psychology at the Slovak Academy of Sciences observed that COVID-19–related conspiracy beliefs were strongly correlated with general conspiratorial thinking and pseudoscientific beliefs, both of which were associated with a low sense of control and decreased trust in institutions. They reported in particular that:

"People who believe in one specific conspiracy theory are very likely to hold other, unrelated, and even contradictory conspiracy beliefs, and they are also far more likely to accept various paranormal or pseudoscientific claims. One study indeed shows that conspiratorial ideation, the tendency to explain complex social and/or political events through conspiracy theories, may be the key factor underlying the rejection of well-established scientific findings."

Spirituality, Religiosity, and "Conspirituality"

It has long been well established that spiritual and religious beliefs provide ways to find or assign meaning to life and its mysteries, and to feel safer in the face of "unknown powers" or "evil forces." Religious systems also fulfill the human need to feel connected to others.

Another role that spirituality might play in the formation of beliefs was proposed by Charlotte Ward and David Voas, who describe the emergence of a new movement they call "Conspirituality." Ward and Voas define conspirituality as a hybrid belief system that merges New Age–type spiritual doctrines with conspiracy theories. Despite important differences between New Age spirituality and conspiratorial thinking, the authors identified several key similarities between them:

- Both are belief systems.
- The process of adherence follows psychological and social motives identified in numerous studies.
- They are built on shared ideas and principles, such as: nothing happens by chance, reality is not what it seems, and simultaneous events must be interconnected.

A Rabbit in the Clouds

Inga Jasinskaja-Lahti and Jolanda Jetten sought to determine whether religiosity, anti-intellectualism, and political distrust could predict belief in conspiracy theories. Their study compared believers (251 participants who self-identified as religious) and non-believers (263 participants who self-identified as non-religious).

The value of their research lies in going beyond the simple question of whether or not a person holds religious beliefs. Their findings revealed that believers and non-believers did not differ significantly in their general level of adherence to conspiracy theories. However, and this is the crucial point, they found that among believers, the importance attributed to their religious worldview was directly associated with stronger belief in conspiracy theories.

In their words:

"Among believers, the importance given to their religious worldview was directly associated with a higher belief in conspiracy theories, and this link was mediated by greater anti-intellectualism and political distrust [...] whereas among non-believers, there was no relationship between a non-religious worldview and belief in conspiracy theories."

They add:

> "We find that it is not the self-categorization as religious or non-religious that matters, but rather the degree of importance attached to the worldview provided by religious beliefs; this is what most strongly predicts people's belief in conspiracy theories."

A Rabbit in the Clouds

Talia Leibovitz and her colleagues examined the factors associated with the likelihood of endorsing COVID-19 conspiracy theories, as well as the mental-health consequences of such beliefs. They surveyed 1,142 participants twice, in April 2020 and May 2020, to detect possible variations.

Their data revealed that about half of the sample (49.7%) believed in at least one conspiracy theory. Leibovitz and colleagues emphasized that their findings provide clear evidence of negative mental-health effects associated with conspiratorial beliefs, particularly increased anxiety. They also observed a connection between religiosity/spirituality and the adoption of conspiracy beliefs.

Thus, religiosity and spirituality can be added to the epistemic, existential, and psychosocial motives previously discussed.

Vukašin Gligorić and colleagues, from the University of Amsterdam (Netherlands), grouped several potential predictors together to determine which factors play a predominant role in the emergence of conspiracy beliefs. When incorporated into a multifactorial model, three factors stood out strongly:

1. Higher levels of spirituality;
2. Lower analytical thinking;
3. Stronger narcissistic traits.

Franks Bradley and Martin W. Bauer, from the Institute of Social Psychology in London, along with Adrian Bangerter from the University of Neuchâtel, analyzed conspiracy theories through the lenses of the cognitive science of religion, social representations theory, and frame theory.

Their analysis led them to propose that conspiracy theories function as "quasi-religious representations", meaning that their content, structure, and function parallel those found in the belief systems of institutionalized religions, even if conspiracy theories do not exhibit all the features of organized religions.

Further Reflections on the Concept of Conspirituality

To add some personal reflections on the concept of conspirituality as defined by Ward and Voas, it can be seen as extremely broad, encompassing a vast range of belief systems and manifesting in diverse forms. It can draw upon many different types of content.

Consider the countless sectarian or religious movements, ideological currents that have fueled revolutions, crusades, inquisitions, religious wars, geopolitical conflicts, terrorist acts, and extremist political ideologies such as Nazism or apartheid. In many of these, one frequently finds a fusion of conspiratorial and spiritual representations.

In most cases, one can easily detect dichotomous concepts, the familiar dualities of good/evil, right/wrong, for/against, chosen/fallen, heaven/hell, faithful/traitor, oppressor/oppressed, curse/blessing, and so on.

A Rabbit in the Clouds

These oppositional frameworks reveal how conspirituality often merges the emotional reassurance of spirituality with the explanatory power of conspiratorial thinking, offering both meaning and control in a complex, uncertain world.

*****

Here is a table summarizing the main motives and social factors involved in the formation and adoption of beliefs.

SOCIAL MOTIVE	EXAMPLE
Need for Uniqueness Linked to self-esteem; desire to stand out and be independent from others.	First Tattoo Suggestions That Are Unique & Meaningful
Existential Motivations Fundamental social drivers in the search for meaning; tied to compensatory control and intentionality bias.	*"As intellectualism suppresses belief in magic, the world's processes become disenchanted, lose the magical significance, and henceforth simply 'are' and 'happen' but no longer signify anything."* MAX WEBER

Causal Attribution Biais Explaining uncontrollable events through deliberate or mystical causes rather than chance.	
Conspirituality Hybrid belief system merging spiritual doctrines with conspiratorial ideas.	

A Rabbit in the Clouds

At the end of this chapter, it is important to emphasize that it is the combination of various factors that determines the strength of a believer's convictions, their worldview, the degree to which these beliefs are integrated into their personal choices, and the depth of their commitment to them.

In the next chapter, we will add to the psychological and social factors the role played by certain personality traits.

CHAPTER 3
PERSONALITY TRAITS

Preamble to This Chapter

A question arises here: Is there a particular and consistent profile shared by those who engage in conspiracy thinking, join religious sects, or associate with groups structured around marginal ideologies? In short, is there a typical "conspiracist personality profile?"

The answer, of course, is not as simple as the question and cannot be reduced to a mere yes or no. As with most human behaviors, the triggering, motivational, and sustaining factors are numerous and vary from one person to another. Many researchers have examined this question, and in this chapter, we will explore what they have discovered regarding the links between "personality traits" and "beliefs."

Personality, Personality Traits, and Personality Disorders: Some Definitions and Clarifications

Let us first distinguish between the three concepts of "personality," "personality trait," and "personality disorder." In psychology, personality refers to a combination of a person's emotional characteristics, attitudes, and behaviors. Personality traits are, in a sense, the "building materials" and "architecture" of the personality. It is important to understand that a personality trait is not the same thing as a personality disorder.

A Rabbit in the Clouds

Dr. Andrew E. Skodol, from the University of Arizona College of Medicine, explains clearly:

"Personality traits are relatively stable patterns of thinking, perceiving, reacting, and relating. For example, some people tend to be moody and closed off, while others tend to be more outgoing and sociable. Personality disorders occur when these traits are so intense, rigid, and maladaptive that they become a source of problems at work, in school, and/or in the person's relationships with others."

For example, if we observe paranoid personality traits in someone, this does not necessarily mean they suffer from a paranoid personality disorder. According to the DSM-5,

"Patients with paranoid personality disorder suspect others of exploiting, deceiving, or harming them. They believe they can be attacked at any time and without reason. They persist in their suspicions and thoughts even in the absence of evidence or on the basis of weak evidence."

It would therefore be a mistake to assume that the presence of paranoid traits automatically means a person has a paranoid disorder. We can all feel "a bit paranoid" at times, imagining, for example, that others are judging us, that a comment in a meeting was aimed at us, or that colleagues change the subject when we approach because they were talking about us. Mild paranoid traits do not necessarily impair one's rationality or self-awareness.

A Rabbit in the Clouds

Personality and Beliefs

Our personality traits play a key role in how we act, and especially how we react, to life events, how we interact with others, and how we function in our social, professional, and intimate relationships. Some traits, more pronounced in certain individuals than in others, may lead to reactions and behaviors that are too intense for some or too mild for others, rigid or flexible, resistant or resigned, impulsive or reflective, and so on.

A large body of research supports the idea that our personality traits play a role in the formation of our beliefs, the importance we attach to them, the way they influence our daily lives, and the way we react when we are called to defend them. For instance, when our beliefs are confronted or challenged, our personality traits affect our reactions, our resistance to opposing arguments and opinions that contradict what we believe.

Regardless of one's beliefs, most people reject the idea of being deceived, gullible, naive, fanatical, extremist, or psychologically unstable. It is important to stress that the information presented here is not binary. It would be a mistake to ask whether we are or are not narcissistic, Machiavellian, or psychopathic, since such a question cannot be answered with a simple "yes" or "no." Motives, factors, and personality traits should be viewed as variables that differ from one individual to another, existing along a continuum of degrees, while remaining relatively stable over time within the same person.

A Rabbit in the Clouds

Although it is essential to remain nuanced when discussing the link between personality and beliefs, it would be equally wrong to overlook it entirely. The question remains highly relevant. A considerable number of researchers have examined the personality–belief connection, identifying interesting predictive patterns, though no single "typical profile" has been established. However, certain specific personality traits can predispose individuals to be, become, or remain more inclined toward belief than others.

Vassilis Saroglou, from the Université de Louvain, reviewed 71 studies conducted in 19 countries, involving a total of 21,715 participants, examining the link between religious beliefs and the five personality traits defined by the "Big Five" model, better known by the acronym O.C.E.A.N.: Openness, Conscientiousness, Extraversion, Agreeableness, and Neuroticism.

The two traits found most frequently among religious individuals were Agreeableness and Conscientiousness. According to Saroglou:

"They are generally more altruistic, compassionate, and cooperative, characteristics linked to agreeableness, and more thoughtful, self-disciplined, and rule-abiding, which correspond to conscientiousness."

In 2019, Andreas Goreis and Martin Voracek from the University of Vienna (Austria) published a meta-analysis on the association between the five O.C.E.A.N. personality factors and conspiracy beliefs. Their study revealed that none of the five Big Five traits were significantly associated with conspiratorial thinking.

Other studies have focused on cognitive styles. Like personality traits, cognitive styles are also relatively stable and refer to the mental processes people use when drawing conclusions, making decisions, forming opinions, or adopting beliefs.

As discussed in Chapter 1, a person's intuitive thinking style (or intuitive cognitive style) has major consequences for how they respond to a wide range of situations and phenomena, notably through reliance on shortcuts, approximate estimations, and broad general impressions at the expense of detailed, analytical reasoning.

Thus, an intuitive cognitive style is a reliable predictor of belief in conspiracy theories, and conversely, conspiracy belief and intuitive thinking share many overlapping correlates.

Pascal Wagner-Egger cites several references showing that a large number of similar predictors have been identified between conspiracy theories, prejudices, stereotypes, and racism.

The Narcissistic Personality Trait

A group of researchers from several European universities has shown that the narcissistic personality trait, that is, a grandiose self-perception combined with low self-esteem, is a strong predictor of adherence to beliefs and conspiracy theories. It is important to note that narcissism always involves interpersonal relationships: a person must compare themselves to others in order to define how unique, special, or superior they are.

A Rabbit in the Clouds

Since beliefs are rarely individual, it is common for people who share them to form groups. These groups can take many forms, from small, informal circles to large, structured, hierarchical organizations. Groups can form around religious beliefs, social class, political allegiance, language, national identity, socioeconomic status, occupation, education level, cultural values, or any other ideological current. Such contexts provide fertile ground for collective narcissism.

In social psychology, collective narcissism refers to the tendency to exaggerate the value and importance of one's own group, combined with the conviction that it is exceptional yet insufficiently recognized or appreciated by others.

Although narcissism primarily applies to individuals, collective narcissism operates on the principle that a group can function as a narcissistic entity. Agnieszka Golec de Zavala and Dorottya Lantos found that collective narcissism is a predictor of conspiracy beliefs that accuse outgroups (including certain minorities) of plotting against their ingroup. In other words, collective narcissism involves a strong ingroup bias, in which members believe that their group has uncovered conspiracies among outsiders, while failing to see any within their own ranks.

A series of studies conducted by Zavala et al. also found that collective narcissism can predict intergroup aggression. This aggression can take many forms, from disdain for those who do not share the group's beliefs to the most extreme forms of fanaticism.

A Rabbit in the Clouds

Once a group of individuals becomes convinced that it holds the truth about a subject, that it is superior to others because it possesses "special" knowledge, and that it has a mission to persuade, correct wrongs, save the world, or fight opponents, the conditions become ripe for fanaticism and extremism.

Sophie Kjærvik and Brad Bushman, from Ohio State University, conducted a meta-analysis combining 437 independent studies involving more than 123,000 participants. They found that narcissism is associated with both aggression and violence. As expected, the link between narcissism and aggression was stronger under provocation than under neutral conditions. Nevertheless, even in the absence of provocation, a significant correlation between narcissism and aggression was observed.

This meta-analysis also revealed that narcissism is linked to multiple forms of aggression, both direct and indirect, physical and verbal, ranging from harassment and intimidation to vandalism and brutality.

Other studies have shown that religious fundamentalism can amplify collective narcissism, thereby increasing the likelihood of radical, extremist, and fanatical behavior toward others.

Once again, when a group, whether religious, ideological, or political, is convinced that it holds the correct worldview, that it possesses exclusive knowledge exposing malevolent powers or dark forces hidden from the majority, and that it has a duty to reveal the truth to the world, such conviction inevitably feeds both individual and collective narcissism.

The Paranoid Personality Test

Roland Imhoff and Pia Lamberty conducted a meta-analysis, which, despite some important nuances, demonstrated a reliable association between paranoia and belief in conspiracy theories.

Anna Greenburgh and Nichola J. Raihani confirmed that paranoia and conspiratorial thinking share many common risk factors (such as victimization and social isolation) and exhibit many similar phenomenological characteristics, including a tendency to attribute malicious intent to external agents.

Greenburgh and Raihani make it clear, however, that it would be inaccurate to claim that individuals with paranoid traits always believe in conspiracy theories, just as it would be wrong to assume that everyone who believes in conspiracy theories is paranoid. It is important to remember that we are talking about paranoid personality traits, not paranoid personality disorder.

Research has shown that paranoid thinking involves idiosyncratic beliefs such as:

- "There are conspiracies directed against me,"
- "Certain people are working together to harm me personally,"
- "All those people over there are watching and judging me."

Furthermore, such conspiratorial beliefs are often shared with others, which can foster collective attitudes with harmful social consequences, particularly the ingroup/outgroup bias (us versus them) and feelings of hostility toward outsiders.

It is worth recalling that although much research has focused on conspiracy theories, the elements discussed here are not limited exclusively to them.

Indeed, Bradley et al., drawing on various perspectives from the cognitive sciences, found that the main features of conspiratorial content are "quasi-religious representations" in the sense that their content, forms, and functions parallel those found in the belief systems of institutionalized religions.

The Dark Triad

Sara Hughes and Laura Machan, from Sheffield Hallam University in the United Kingdom, extended research on personality traits and the tendency to adopt conspiratorial beliefs. They observed the prevalence of what they call the "Dark Triad," composed of psychopathic traits, Machiavellianism, and collective narcissism, and sought to validate these findings.

More specifically, their research confirmed what has often been observed: collective narcissism is built upon the conviction of possessing information unknown to others, that this "privileged knowledge" fuels the need for uniqueness, stimulates a sense of superiority, and even fosters the belief in being entrusted with a special mission.

A Rabbit in the Clouds

Machiavellianism refers to a lack of empathy toward others and a manipulative tendency aimed at achieving one's goals. It involves a variety of strategies used to self-justify one's beliefs or theories as valid, to view one's motives for spreading them as legitimate, and to see one's methods as justified. In short, Machiavellianism largely rests on the maxim: "the end justifies the means."

Psychopathic traits are generally divided into two dimensions:

1. Primary psychopathy, characterized by emotional coldness and lack of empathy;

2. Secondary psychopathy, characterized by impulsivity, unscrupulous use of lies, and absence of remorse.

Nuances and Clarifications

Here are a few important nuances and clarifications provided by Poupart and Bouscail:

- Conspiratorial thinking is not a psychological disorder. It "belongs much more to the domains of political discourse and the social and cognitive sciences than to psychiatry."

- Believing in a conspiracy theory is not symptomatic of a mental disorder or a source of psychological suffering or maladjustment in daily life.

Nevertheless, Poupart and Bouscail refer to a recent literature review on the psychological research concerning conspiracy theories, which reveals that:

- Empirical studies "have observed a significant link between adherence to conspiratorial beliefs and paranoid or schizotypal personality traits, as well as delusional ideation in non-clinical subjects."

- Furthermore, Poupart and Bouscail add that their clinical experience has led them to observe that "conspiratorial thinking in certain patients fits within the broad psychopathological spectrum of narcissistic–identity suffering and borderline functioning."

- This observation raises questions for them about "the role of conspiratorial thinking in maintaining the psychological balance of such patients."

Much more research will need to be conducted in this field, and it is reasonable to expect that we will benefit from its findings in the years to come.

*****

At the end of this first part, I propose a summary using as an example the flying objects observed and shot down by the U.S. and Canadian militaries in February 2023. Beyond the "sensational" nature of this news, many elements remained unknown at the time these events occurred. The stage was therefore perfectly set for belief-makers and theorists of all kinds to get to work.

We will limit ourselves here to comparing three popular belief systems that offered their own interpretations of this news and the mysteries surrounding it:

A Rabbit in the Clouds

- the Roswell incident,

- the beliefs of the followers of Raël, and

- the beliefs of the followers of Project Blue Beam.

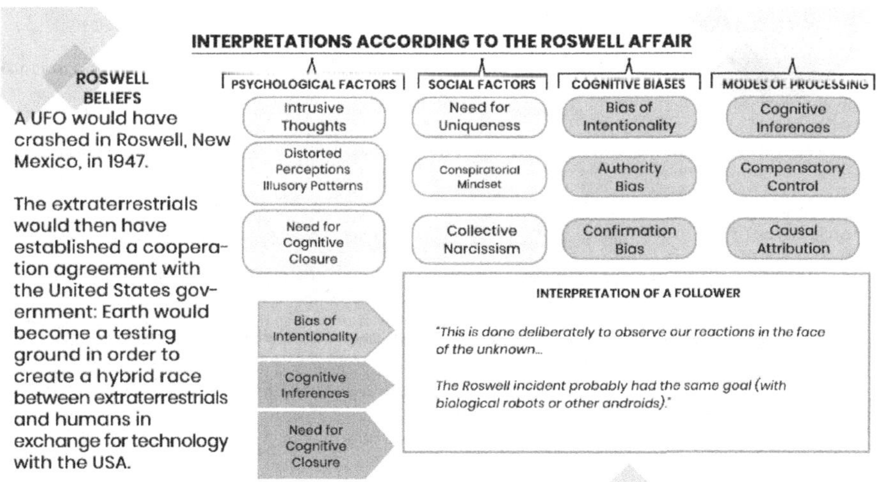

A Rabbit in the Clouds

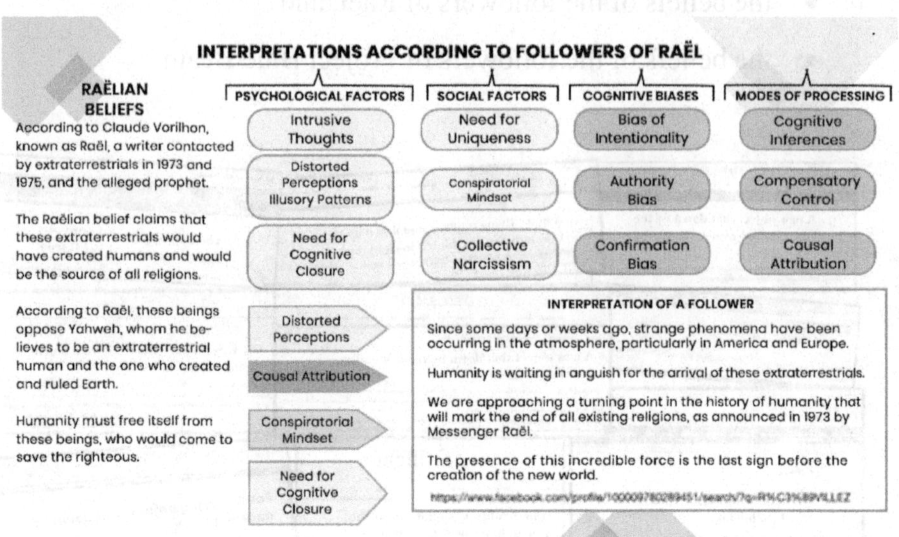

According to Claude Vorilhon, known as Raël, a writer contacted by extraterrestrials in 1973 and 1975, and the alleged prophet.

The Raëlian belief claims that these extraterrestrials would have created humans and would be the source of all religions.

According to Raël, these beings oppose Yahweh, whom he believes to be an extraterrestrial human and the one who created and ruled Earth.

Humanity must free itself from these beings, who would come to save the righteous.

INTERPRETATIONS ACCORDING TO FOLLOWERS OF THE BLUE BEAM PROJECT

BELIEFS OF THE BLUE BEAM PROJECT

In the year 1990, according to researcher Serge Monast, the Blue Beam Project aims to create an invasion by extraterrestrials in order to control humanity through a terrible psychological shock: "to simulate an extraterrestrial invasion by holograms in order to implant a new religion: to destroy all existing religions in order to establish a single and unique one."

PSYCHOLOGICAL FACTORS	SOCIAL FACTORS	COGNITIVE BIASES	MODES OF PROCESSING
Intrusive Thoughts	Need for Uniqueness	Bias of Intentionality	Cognitive Inferences
Distorted Perceptions Illusory Patterns	Conspiratorial Mindset	Authority Bias	Compensatory Control
Need for Cognitive Closure	Collective Narcissism	Confirmation Bias	Causal Attribution

Need for Cognitive Closure

Bias of Intentionality

Cognitive Inferences

Compensatory Control

INTERPRETATION ACCORDING TO THE BLUE BEAM PROJECT

Blue Beam Project February 3, 2023

The Blue Beam Project claims that the extraterrestrial invasion will occur globally and that humanity will face an extraterrestrial invasion that will occur in all major cities simultaneously. The shock and panic generated will force people, even the most desperate, to accept the New World Order.

PART TWO

ADHERENCE AND COMMITMENT TO A BELIEF SYSTEM

CHAPTER 4
THE ESCALATION OF COMMITMENT: FROM HARMLESS BELIEF TO BELLIGERENT FUNDAMENTALISM

Have you ever noticed how difficult it is for a human being, once committed to a path that turns out to be a mistake, to admit it, question themselves, and turn back? This phenomenon has been observed in countless situations. It manifests itself just as much during friendly discussions, where valid and relevant arguments seem to have no effect on anyone's firmly held position, as in political decisions, financial risk-taking, or ideological stances and opinions.

One of the classic examples is known as the "Concorde Effect." It refers to the stubbornness of the French and British governments in continuing the Concorde project, into which enormous sums of money had been invested. Yet, as early as 1973, experts already knew that its commercial operation would never be profitable.

Barry M. Staw, a researcher in organizational psychology at the University of California, Berkeley, became interested in the behavior of financial investors and how they act after making bad decisions. He observed a pattern called "escalation of commitment."

This occurs when an individual, organization, or society faces increasingly negative outcomes following a decision, project launch, or financial investment, yet continues down the same path, rather than changing direction. The grip of this cycle is so strong that those caught in it desperately persist in maintaining irrational behaviors, which are mostly anchored in their past positions, decisions, or actions.

A Universal Mechanism

Understanding that this mechanism is universal is essential because it manifests everywhere and in everything, across all domains, cultures, and societies, at every level and in every era. Escalation of commitment can occur for both good and bad reasons, in all sectors, from charitable organizations to criminal networks, community services, and even state imperialism.

Although the consequences are not always of the same magnitude, and some may be minor or even insignificant, the underlying mechanisms remain remarkably similar.

No one is immune to the risk of getting caught in this cycle of commitment, once they begin to defend and justify their own actions, projects, opinions, beliefs, or convictions.

The Determinants of Escalation of Commitment

The determinants of escalation of commitment have been studied and classified into four categories by Isabelle Royer:

1. Psychological determinants
2. Sociological determinants
3. Organizational determinants
4. Project-related characteristics

1. Psychological Determinants

These involve all the psychological factors that influence an individual. Among them are cognitive biases, through which people selectively focus on facts, information, and evidence that confirm their ideas, beliefs, and decisions, while ignoring or minimizing anything that contradicts them.

As you have likely noticed, cognitive biases play a major role in a wide range of circumstances. Some affect the way individuals process information, while others act directly on their persistence in commitment, helping them to psychologically convince themselves of the correctness of their beliefs, decisions, or actions.

In short, psychological determinants act like lenses that distort reality to make it fit one's beliefs, allowing individuals to self-persuade that their previous decisions were the right ones, and to convince themselves that it is justified to continue in the same direction.

2. Sociological Determinants

These stem from external pressures acting on the individual, pushing them to behave according to established norms or rules. They include all the social incentives that encourage one to persist in an action, project, or belief system.

The anticipation, or even illusion, of a positive outcome is a major obstacle to giving up a project or belief, since withdrawing would mean admitting an error or failure, and risking social embarrassment.

A Rabbit in the Clouds

Conversely, those who persist in their commitment often hope to benefit from the "hero effect", a social admiration for people who persevere despite adversity and eventually triumph. We see this, for instance, in literature and cinema, which frequently celebrate as heroes or heroines historical figures who, despite overwhelming obstacles, demonstrate perseverance and resilience, and ultimately succeed against all odds.

When an individual is committed to an action doomed to fail or to a belief that is objectively challenged, it becomes very difficult for them to admit and reveal their mistakes, for fear of damaging their self-image. They cling to those beliefs as if they were part of their own identity.

3. Organizational Determinants

These arise from the organization behind the decision, project, or belief system. Generally, once a decision is made, it triggers a series of related decisions. The more institutionalized the beliefs and actions become, the more they are accepted and reinforced, and the less likely they are to be questioned. They tend to self-perpetuate.

If, however, these beliefs or actions are criticized or challenged, coercive mechanisms may come into play to pressure dissenters into conformity, to make them think, believe, and act in line with the organization's imposed framework.

As you can see, organizational determinants are closely linked to the sociological determinants discussed above.

4. Project-Related Determinants

These concern the intrinsic structure of the project itself, which is often built upon certain beliefs and supported by enticing promises. The anticipation of such rewards reinforces the escalation of commitment.

If, unfortunately, obstacles arise during the course of the project, such as unforeseen expenses, unfulfilled promises, unexpected setbacks, or even signs of fraud, the initial motivations that were driven by short-term achievement tend to transform into expectations of future results.

In other words, rather than accepting failure, people double down, continuing to invest emotionally, financially, or ideologically in the hope that eventual success will vindicate their original choice.

The Opponent-Process Theory of Motivation

Let me tell you about an experiment I have conducted several times with my psychology students to help them understand this fascinating process. The exercise is a bidding game for a five-dollar bill. Each bid increases by 25 cents, and the bill will go to the highest bidder. However, there's a catch that participants don't immediately notice. I state the rule quickly as follows:

"The highest bidder will get the $5 bill for the amount they bid. However, the second-highest bidder must also pay me the amount they bid."

A Rabbit in the Clouds

This means that for the person who comes in second place, their bid becomes a net loss. Without giving them much time to think, I start the auction. Motivated by the lure of gain, the boldest participants begin with bids of 25¢, 50¢, 75¢, $1, $1.25, and so on. As the bidding continues, some players drop out until only two remain.

Then comes the turning point: when one of the last two decides to give up and let the other get the $5 bill at a "discount." At that moment, I remind them:

"Remember, the highest bidder gets the $5 bill for their bid, but the runner-up still has to pay me their last bid."

Suddenly, motivations shift entirely. The second-highest bidder realizes they will lose money with no benefit, and jumps back into the bidding. As you might guess, the two remaining bidders continue to outbid each other until the offers exceed $5.

Their initial motivation, to buy a $5 bill at a discount, has reversed into a desire to minimize their loss. What makes this example so striking is that the final bids often go far beyond $5. Each of the last two participants wants "at all costs" to be the one who wins the $5 bill in order to reduce their loss.

After a few bids surpass $5, I end the auction and tell them they don't owe me anything. The point of the exercise is to help them experience the opponent-process theory of motivation.

Outside this mischievous little game, the opponent-process theory of motivation appears frequently in everyday life. Consider, for instance, a beginner who goes to a casino for the first time and finds it fun to wager a few dollars on a slot machine flashing bells, sevens, and lemons to cheerful music. But two insidious traps await.

The first trap lies in occasional wins. After losing several dollars, when the player wins a small amount, their perception of reality becomes distorted, they no longer accurately distinguish between "losing" and "winning." If, for example, they lose $20 or $30 over ten or twenty plays and then win $2, the momentary thrill amplifies their euphoria, clouding their judgment. They believe this small win proves they should keep playing.

The second trap is that "the more they lose, the more they keep betting," believing they will win back their losses. And since occasional wins do happen in gambling, the desire to continue becomes self-reinforcing.

Thus, the person's initial motivation (to have fun, play, and maybe win money) gradually transforms into an opposite motivation (an antagonist process), where their compulsive play is driven by the need to relieve the discomfort caused by their losses.

We owe the opponent-process theory of motivation to Richard Solomon and John Corbit. This theory has inspired extensive research into addictive behaviors, including those related to psychoactive substances, gambling, high-risk activities, and even toxic human relationships.

A Rabbit in the Clouds

In the case of drug use, for example, the classic pattern is as follows: an individual experiences intense pleasure during their early use. However, with repeated consumption, an adaptation mechanism (tolerance) develops, meaning they must increase the dose or frequency to achieve the same euphoric effects. When the drug's effects wear off, withdrawal symptoms emerge, often extremely unpleasant. This can lead to a vicious cycle, where the person's motivation shifts from seeking pleasure to avoiding the pain and distress of withdrawal.

Researchers George F. Koob and Michel Le Moal have studied the neurobiological mechanisms of addiction. They have shown how chronic substance use can disrupt certain brain structures, including those that produce pleasure-related neurotransmitters such as dopamine and opioid peptides.

Moreover, abstinence and withdrawal lead to a surge of stress-related neurotransmitters and hormones, including cortisol, adrenaline, corticosteroids, and noradrenaline. This progressive loss of pleasure, combined with the distressing emotional and physical effects of withdrawal, drives the individual to use again, not to feel good anymore, but to stop feeling bad.

Opponent Processes of Motivation and Escalation of Commitment

The opponent-process theory of motivation, formulated by Solomon and Corbit, and Staw's theory of escalation of commitment, both apply to a wide range of human domains, including, of course, the adherence to and defense of one's ideas, opinions, and beliefs, as we will see in the next chapter.

Let us return to the example of a discussion among friends, where arguments and counterarguments on a controversial topic are flying in all directions. The tone rises, emotions flare, and certain statements trigger strong reactions. The longer the discussion continues, the more each person becomes entrenched in their position, to the point where sound and reasonable ideas are dismissed outright simply because they contradict one's own stance. It goes without saying that such situations have often broken friendships and families, and even led to political and religious conflicts and wars.

As Léa Fensch-Carrat observed, "the more an individual invests resources (time, money, etc.) in a cause, the less likely they are to change their mind about it." That is why those who adhere to a belief system and commit themselves to promoting or defending it become increasingly resistant to anything that might shake that system, rejecting arguments or information that could generate doubt or self-questioning.

This is known as the "absurd trap" (le piège abscons), a situation in which a person stubbornly persists in a strategy, decision, or belief into which they have already invested a great deal of time, money, energy, and emotion.

The same absurd trap that ensnares the drug user or the casino gambler also lies in wait for the believer. It is one of the core ingredients of the escalation of commitment within a belief system, pulling the individual into a genuine vicious cycle.

It usually begins with a cluster of appealing ideas or theories that attempt to explain mysterious, ambiguous, or obscure topics, or that offer comforting answers to existential anxieties. Some of these ideas may resonate with the individual's values and upbringing (freedom, health, respect, happiness, etc.), producing a soothing effect.

In addition, the need for uniqueness, discussed earlier, reinforces the feeling of possessing privileged information that perfectly fits one's intuitive ways of thinking. It is not uncommon for a person to feel personally "called", as if by a higher mission, to take action and invest themselves in a noble cause or extraordinary mission, such as revealing the Truth, spreading the Good News, exposing the Conspiracy, or unmasking Corruption.

There is also the promise of reward, whether material, financial, moral, or supernatural. For some, the reward is money; for others, it is heavenly salvation; and for still others, it is the pride of serving a great humanitarian cause, of contributing to a "promised land."

However, at some point, the follower inevitably encounters opposition, contradictory information, overwhelming evidence that challenges their beliefs, opposition from friends, relatives, or partners, negative judgments, or even threats of rejection. This adversity provokes a range of unpleasant emotions, stress, and inner discomfort.

A Rabbit in the Clouds

Each step, each argument, each debate, and each confrontation ends up reinforcing the individual's beliefs. The opponent processes and escalation of commitment make the person highly resistant to ideas that contradict their convictions. And, much like the casino player, their motivation shifts over time. The well-being once produced by the belief system and by the commitment to spreading it becomes intertwined with discomfort and tension brought on by opposition.

The very idea of being wrong becomes so uncomfortable that it fuels an even stronger need to reinforce and defend one's beliefs. And it is precisely at this stage that numerous cognitive biases come into play, reinforcing the illusion of certainty and protecting the fragile balance of conviction.

Cognitive Dissonance

In the mid-20th century, psychologist Leon Festinger of the Massachusetts Institute of Technology (MIT) sought to understand how humans react when facts contradict their deeply held beliefs.

To study this phenomenon, Festinger infiltrated *The Seekers*, a religious sect based in the Chicago area. Its founder, Dorothy Martin, also known as *Sister Thedra*, claimed to communicate with extraterrestrials who had told her that the end of the world would occur on December 21, 1954. Naturally, only the group's faithful members would be saved by boarding a spaceship. Some followers were so convinced that they sold their possessions, quit their jobs, and prepared for their rescue by aliens.

A Rabbit in the Clouds

Festinger was curious to observe how the believers would react once they realized that this prophecy was false, that the end of the world would not occur as Sister Thedra had predicted.

Indeed, on the night of December 21, 1954, the faith of *The Seekers* was put to the test by an undeniable reality: no apocalypse, no spaceship, no interstellar journey. Disoriented and shocked by the deception, some members quickly left the sect. Yet, many others stayed.

But how could they remain so naive and credulous? Why didn't they all leave once the fraud was clear? One possible explanation lies in a "new revelation" from Sister Thedra. In the early morning hours, she announced to the group that she had received a new message from the extraterrestrials: their prayers had been heard, and thanks to them, the end of the world had been averted. The Seekers, she claimed, had saved the planet from destruction.

This spontaneous reinterpretation of their doctrine not only reinforced their faith but also deepened their commitment to it. Whereas the Seekers had previously shunned the media and refrained from proselytizing, from that day forward, they contacted journalists to share their "miraculous news" and launched a major recruitment campaign. It was following this extraordinary observation that Festinger formulated the theory of cognitive dissonance.

A Rabbit in the Clouds

What Is Cognitive Dissonance?

Cognitive dissonance refers to the psychological discomfort and tension experienced by a person when:

- Their ideas, beliefs, values, behaviors, or emotions are inconsistent or contradictory;

- Their actions do not align with their values or principles.

- Or when external, undeniable facts invalidate their beliefs or theories.

When cognitive dissonance occurs, the resulting psychological distress can manifest as disappointment, shame, guilt, embarrassment, anxiety, or anger.

To reduce this discomfort, individuals seek ways to restore internal harmony. They may:

- Change their behavior by abandoning the belief or adopting a new one;

- Reinterpret or modify their beliefs to make them fit the facts;

- Or downplay or ignore the importance of disconfirming evidence.

The Seekers' Example

In the case of *The Seekers*, the most effective way to relieve their dissonance would have been to question their beliefs, confront their prophetess, or leave the group altogether.

A Rabbit in the Clouds

However, such an action is emotionally difficult because it requires admitting one's mistake, sometimes after months or years of devotion, and accepting that one has been deceived or manipulated. Moreover, leaving the group can mean severing strong emotional and social bonds formed within it.

For these reasons and others, escaping a belief system in which one has invested time, affection, money, and relationships can be extraordinarily challenging.

The other way to reduce dissonance is the opposite: to dive deeper into the belief system, falling into the "absurd trap" by intensifying one's commitment, even if it means altering, modifying, or adding new elements to the original belief system in order to preserve its internal coherence.

CHAPTER 5
LIES, FAKE NEWS, RUMORS, AND PROPAGANDA

We are bombarded daily with information, and a certain percentage of it consists of false news (fake news), misleading claims, dishonest advertising, deceptive political propaganda, and presumptuous rumors, among others.

Let us be clear: lying is part of human life. No one can truthfully claim to have never lied, whether to a child, an employer, a partner, or a spouse, or to have never exaggerated, embellished a story, or told a white lie.

Yet, there is a vast difference between telling a fantastic story to a child or using a euphemism to spare someone's feelings and deliberately deceiving another person to extract money, sexual favors, or political votes. One of humanity's great paradoxes lies in its fascination with lies and its indignation toward them.

Fake news, rumors, and deceitful propaganda have always existed, but today, with the rise of the Internet and social media, coupled with algorithms and artificial intelligence, their creation and spread have reached a level likely unprecedented in human history.

Reality confronts us once again with one of the oddities of human nature: at times, we are victims of lies and fake news, while at other times, we consume them almost compulsively.

A Rabbit in the Clouds

The Spread of Fake News

To understand how false news spreads, Soroush Vosoughi, Deb Roy, and Sinan Aral conducted a large-scale study on the differential diffusion of true and false news shared on Twitter between 2006 and 2017. The dataset included about 126,000 news stories tweeted by 3 million people, amounting to more than 4.5 million retweets.

The classification of news as "true" or "false" was carried out by six independent fact-checking organizations, with a 95–98% agreement rate. Their findings speak volumes about the public's attraction to misinformation.

The study revealed that true news rarely spread beyond 1,000 people, while false news traveled much farther, sometimes reaching up to 100,000 users. In fact, the likelihood of a false story being retweeted was 70% higher than that of a true one.

This extensive research by Vosoughi and colleagues demonstrated that fake news spreads faster and more widely than factual information, across every category of content.

How We Judge Truth and Falsehood

When trying to determine whether a piece of information is true or false, we generally rely on inference; that is, we base our conclusions on certain premises we believe to be valid.

For example, imagine you are sitting on your balcony and hear what sounds like human screams. Almost simultaneously, you see two people talking a few meters away. Because of the distance and surrounding noise, you can't hear what they are saying. However, their gestures and body language give the impression of a heated discussion. You then conclude that these two people must be the source of the screams.

You've reached this conclusion based on three factual premises:

- You heard screams that sounded human.
- You saw two people talking nearby.
- Their gestures appeared animated.

Everything seems logical and consistent. Even without verifying, you sincerely believe your conclusion reflects reality.

If you happen to be wrong, the consequences are minor and inconsequential in this case. But not all our inferences are so harmless. In many situations, invalid inferences can lead to serious and far-reaching consequences.

As we have emphasized several times, there exists a vast number of beliefs and theories covering an enormous range of topics. However, the number of those that have been proven true is significantly smaller.

One major problem is that, beyond objective proof of truth or falsehood, what often dominates is what we believe to be true, our subjective perception of truth, rather than truth itself.

A Rabbit in the Clouds

How Do People Distinguish Truth from Falsehood?

According to Nadia Brashier and Elizabeth J. Marsh of Duke University, judgments about truth are generally based on three types of inferences:

1. Inferring Truth from Base Rates

When faced with an ambiguous or unfamiliar statement, we might assume that people are equally likely to consider it true or false. However, in reality, we tend to believe most of the information we encounter. This is because, in general, most of it is mundane, factual, and easily verifiable.

Brashier and Marsh point out that it is usually reasonable to assume information is true, and if later evidence disproves it, we can revise or reject it.

However, they note that we rarely engage in thorough fact-checking, as this process is time-consuming and cognitively demanding.

2. Inferring Truth from Feelings

We also tend to infer truth based on how we feel about the information. When a message is easy to process, because it is simple, familiar, or fluently worded, it is more likely to be perceived as true. Moreover, repetition strongly reinforces this perception. The more often we hear or see a piece of information, the more we feel that we "know" it, and therefore, the more true it seems. This cognitive shortcut allows us to avoid the mental effort of verification.

A Rabbit in the Clouds

Numerous studies have found similar results: whether through traditional media or social networks, repeated exposure to the same information makes people more likely to believe it, while underestimating its source, credibility, and authenticity. In short, the more often we hear something, the truer it becomes in our minds, a phenomenon known as the illusory truth effect. Matthew L. Stanley and his colleagues tested whether this effect occurs even when people claim not to pay attention to the information around them.

Across three experiments, they found that even when participants' minds wandered, repetition still increased their truth judgments. This means that the illusory truth effect happens even when we're distracted or disengaged. As Marie-Ève Carignan and her team emphasize, "'Truth' can be a fluid notion in many cases. Beyond the diversity of perspectives, 'information' or 'facts' can never fully capture the complexity of the reality they seek to describe."

3. Inferring Truth from Consistency with Memory

In general, we rely on our existing knowledge or beliefs to interpret the world, resolve ambiguities, and predict future events. We use what we already know (or believe we know) to judge whether new information is true or false.

Thus, when new information fits well with our prior beliefs, we are more likely to consider it true.

However, the problem is that many of our "known facts" are inaccurate, incomplete, or outright false, and we often confuse opinions or assumptions with established facts.

We may even hold strong convictions about things that are, at present, impossible to know for certain.

While it is generally possible to verify or update our knowledge when confronted with new information, Brashier and Marsh note that very few people actually engage in this effortful process of comparison and correction.

Supporting Evidence from Other Researchers

The three inference types proposed by Brashier and Marsh are supported by numerous studies.

For instance, Gordon Pennycook and David G. Rand reviewed 14 studies in which participants were asked to evaluate the accuracy of political messages on social media.

The researchers presented an equal number of true and false headlines in random order and found that participants were better at distinguishing truth from falsehood when the content aligned with their political beliefs.

When it conflicted with their views, their discernment declined. Pennycook and Rand suggest that these errors stem from automatic, intuitive, and incomplete mental processes.

A Rabbit in the Clouds

Similarly, psychologists Anuj K. Shah and Daniel M. Oppenheimer argue that heuristics (mental shortcuts) reduce our ability to judge truth effectively by:

- Examining fewer cues;
- Considering fewer alternatives;
- Integrating less information into the evaluation process.
- And giving more weight to familiar, easy-to-process cues.

The Consequences

Brashier and Marsh conclude that relying on these heuristics as a way to detect misinformation, fake news, rumors, fabrications, or outright lies can lead to serious consequences, such as:

- Refusing to vaccinate children out of fear of autism;
- Turning to charlatans or superstitious rituals instead of medical treatment;
- Arming oneself against imaginary threats;
- Or encouraging violence and hostility based on unverified rumors and assumptions.

In short, our intuitive ways of inferring truth, though often efficient, make us vulnerable to deception in an era of overwhelming information.

Why Are People So Attracted to Fake News?

Jonah Berger and Katherine L. Milkman sought to understand why certain online content spreads more widely than others.

A Rabbit in the Clouds

To answer this question, they examined the relationship between the emotions evoked by news content and its virality. Analyzing 7,000 New York Times articles published over three months, they found that stories that trigger stronger physiological arousal linked to emotions such as fear, anger, surprise, or joy were more likely to go viral.

In contrast, content that evokes soothing or low-arousal emotions (e.g., sadness, boredom) was significantly less likely to spread.

Brady, Gantman, and Van Bavel explored whether moral and emotional content captures more attention than neutral content, and whether this explains why such messages tend to go viral more often.

Through a series of controlled lab experiments analyzing nearly 50,000 politically themed tweets, they discovered that moral and emotional content captured early visual attention far more effectively than neutral messages.

This increased attentional capture was strongly correlated with a higher number of retweets in political discussions and online debates.

Mufan Luo, Jeffrey T. Hancock, and David M. Markowitz investigated whether "approval cues", such as the number of "likes" on Facebook, affect people's perception of credibility and ability to detect false information.

Their results showed that headlines with many "likes" were judged as more credible, which improved people's ability to identify true news, but also reduced their accuracy in detecting fake news.

A Rabbit in the Clouds

These findings support Timothy Levine's Truth-Default Theory (TDT), which posits that:

- Honesty is the default mode of human communication.
- People are truthful unless they have a motive to deceive.
- People believe others by default, unless given a reason to be suspicious or doubtful.
- This truth-default state makes humans vulnerable to false and misleading information.
- Individuals tend to accept information that aligns with their preexisting beliefs and is prevalent in their communication environment.

In short, and at the risk of stating the obvious, human beings are highly suggestible, especially when they are shown where and what to look at in order to "see the rabbit in the clouds."

And, as Pascal Wagner-Egger aptly puts it:

"It takes far more time and energy to refute false beliefs or information than to produce them."

CHAPTER 6
AN EXAMPLE OF BELIEF-MAKING

The More Things Change, the More They Stay the Same!

A particularly interesting and well-documented study compared the COVID-19 pandemic to three other major pandemics: the Black Death of 1347, the New World smallpox epidemics of 1520, and the Spanish flu pandemic of 1918. The goal of Patterson and his colleagues was to compare these four pandemics to identify similarities and differences between the pathogens, the social and medical context, human reactions and behaviors, as well as the long-term social and economic impact.

Many very relevant findings emerged. Those that interest us most in the context of this chapter concern psychosocial reactions and behaviors. The researchers noted that public reactions during an epidemic or a pandemic have hardly changed in almost any respect: disbelief regarding the disease (its existence, its virulence, its causes, etc.), the spread of misinformation, the lack of clarity in public messages, the resistance and popular opposition to the sanitary measures proposed or imposed, the mistaken evaluations of the severity of the disease, and the individual risk behaviors.

In every epidemic, contradictory information about the disease and its evolution is disseminated (whether voluntarily or not) in various ways, sometimes even at the request of government authorities. Patterson et al. mention an example of censorship in which the Italian government forced a Milan newspaper to stop publishing the daily number of deaths during the Spanish flu because it was too demoralizing for the population. At the same time, and for similar reasons, in the United States, certain public health officials did not disclose the extent of the spread of the epidemic and minimized the dangers it represented. This withholding of information, intended to maintain public morale, weakened the population's trust in public institutions.

Misinformation transmitted by various popular movements also provoked many reactions. It often took weeks and numerous victims before the population finally recognized the seriousness of the epidemics. Denial and popular resistance were often enormous. In 1630 in Italy, doctors were regularly insulted in the streets for warning the public that the Black Death was about to strike.

Closer to our time, at the end of March 2020, some Twitter users began spreading the hashtag #FilmYourHospital on social media. A study on the spread of conspiracy theories on social media showed that this campaign, led by experienced conservative activists, quickly went viral. This propaganda sought to encourage people to go to their local hospitals to take photos and videos in order to prove that the COVID-19 pandemic was a hoax. The premise underlying their conspiratorial belief was that if hospital parking lots and waiting rooms were empty, this would prove that the pandemic was not real, or at least not as serious as reported by health authorities and the media.

A Rabbit in the Clouds

In addition, the commotion caused by the lack of information, the contradictory information, and misinformation generated uncertainty, distrust, and despair, which may have led people to resort to dubious and unreasonable protective methods during epidemics and pandemics. In the 1300s, physicians had no effective remedies, which opened the door to charlatans of all kinds offering their "magical" elixirs to ward off the Black Death. The same was true during the Spanish flu, when quacks urged people to wear camphor bags, or when a certain Dr. Folley claimed that the flu was a "cousin of the plague" and that injections of Yersin's anti-plague serum could cure those suffering from the pandemic flu. We witnessed similar reactions at the beginning of the COVID-19 pandemic, when many of our contemporaries believed they could protect themselves from SARS-CoV-2 with zinc lozenges, colloidal silver, vitamin supplements, and unauthorized medications.

Each epidemic also saw its own resistance movements. As we saw during COVID-19, anti-mask groups had also formed in the United States during the Spanish flu, invoking the lack of scientific evidence of its effectiveness and the violation of constitutional rights.

A glance at the table below shows the similarity of popular reactions during the 1885 smallpox epidemic and the 2020 COVID-19 pandemic.

Source: La Presse, September 29, 1885
(Source: Bibliothèque et Archives du Québec).

Source: La Presse, 19 December 2021

Émeute du 28 septembre 1885 à Montréal. (Source : Bibliothèque et Archives nationales du Québec.)

Manifestation de janvier 2022 à Ottawa. (Source : https://nouvelleslaurentides.ca/le-convoi-de-la-liberte-attire-les-foules-a-ottawa/)

A Rabbit in the Clouds

Drawing by Robert Harris (Public domain image)

Arrests, Ottawa, February 2022. (Courtesy).

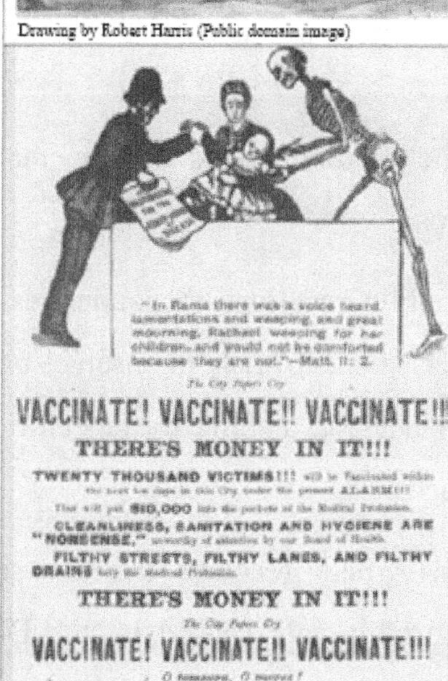

Source :
https://www.museedelhistoire.ca/blog/vaccination-obligatoire/ (Consultée le 30 novembre 2022).

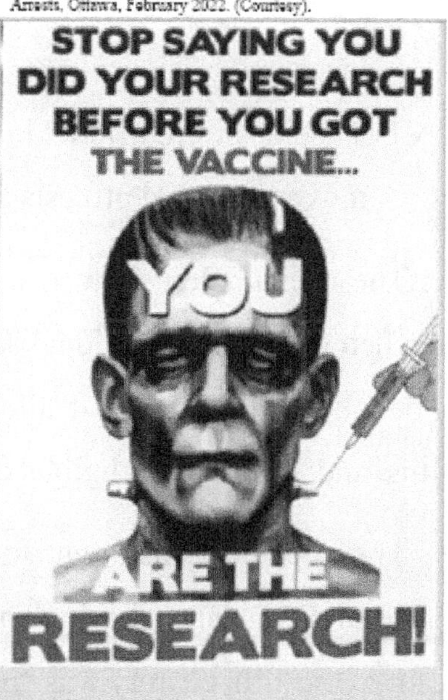

Source : Google image

A Rabbit in the Clouds

The images behind refer to two similar events that occurred 137 years apart. In both cases:

- The virus spread more or less rapidly depending on the affected areas and regions.

- There were a significant number of hospitalizations, infections, and deaths among the most vulnerable, poor people, and those living in poor sanitary conditions.

- Isolation, quarantine of infected individuals, and hospitalization of the sick were recommended, and there were public protests and demonstrations against these health measures.

- Vaccines were offered to the population, and anti-vaccination movements and protests arose.

These examples show that public reactions to pandemics are often determined by individual assessments of risk, themselves based on beliefs of all kinds and allegiances, extending to superstitions, myths, and, of course, religious doctrines.

The medieval historian John Arbeth recounts that for some European Christians, the high mortality rate caused by the Black Death of the 14th century was "proof of God's just judgment" against a sinful and corrupt humanity. The clergy maintained that the world was on the brink of an apocalypse or the reign of the Antichrist. For many believers, it was necessary to take refuge in prayers and processions and to call upon God and saints, particularly known for their mercy or their power against the plague.

A Rabbit in the Clouds

In 1885, the Catholic clergy claimed that the smallpox epidemic striking the population was a demonstration of divine wrath, and that if it afflicted Montreal, it was the fault of a carnival held that February.

The same applied to religious conservatives for whom diseases (including COVID-19) were "God's punishment for our sinful acts." For some religious communities, vaccines were believed to contain human fetal tissue from abortions.

Conspiracy and pseudoscientific theories naturally sprang up everywhere and in all directions. For English speakers, the smallpox epidemic was the fault of French Canadians, who were blamed for the cause. Conversely, for French speakers, smallpox vaccines were an invention by English Canadians to poison Franco-Canadians. Advocates of homeopathy, for their part, considered vaccination a form of "quackery" meant to cause "the poisoning of our children."

For some, the coronavirus was a Chinese biological weapon created in a Wuhan laboratory; for others, it was commissioned by the CIA; and still others believed the pandemic was a population control program created by the Pirbright Institute and Bill Gates.

Denis Goulet, historian of medicine at the University of Montreal, examined the variety of theories that seek to explain the origins and nature of infectious and contagious diseases, a few examples of which include:

- Magico-religious models and members of the clergy who claim it is a severe warning from God to a community that has "set aside His commandments";

- Prayers and religious services directed to invoke protective saints;

- The search for scapegoats, identifying and persecuting them as enemies;

- Blaming political authorities, accusing them of mismanaging the crisis or paving the way for the epidemic (for example, by allowing immigrants to enter);

- Social division: Some members of the working and poor classes suspect the ruling class of wanting to exterminate them, while the latter blames the lower classes for the plague, citing their "unsanitary, primitive, or immoral" living conditions.

- Tensions erupt in most countries, with mutual suspicion and accusations flying on all sides.

Florent Poupart and Manon Bouscail, from the University of Toulouse in France, provide further excellent examples drawn from history:

"Major crises have always been conducive to the production and spread of conspiracy theories. We can observe their resurgence as early as the 11th century in millenarian sects, during the French Revolution at the end of the 18th century, or more recently after the attacks of September 11, 2001. Collective feelings of uncertainty, anxiety, and loss of control, aroused by the collapse of reference points and the questioning of institutions, create a need to make sense of what is being experienced. One of the main functions of these theories is precisely to give shape to chaos, to make it representable and thinkable, by drastically simplifying reality, reducing the number of actors, stakes, factors, variables, and forces at play. In conspiratorial thinking, history is reduced to a succession of plots carried out by a small, all-powerful, all-evil group conspiring cynically against the good and weak people."

The few examples presented in this chapter show how psychological and social motives play a role in the creation of beliefs, and how these beliefs influence emotional and behavioral reactions. Specifically, crises, disasters, and social instability cause anxiety, feelings of uncertainty, and loss of control among individuals. Research on the perception of illusory patterns and the need for cognitive closure shows that our brains are skilled at intuitive, heuristic, quick, and simplistic interpretations. Added to this is a wide range of cognitive biases: the intentionality bias, which leads people to believe in deliberate wills and intentions behind chaos and confusion; the confirmation bias, which influences believers to select and favor information that confirms their beliefs and ignore what might challenge them; and the causal attribution bias, which drives our brains to perceive and even imagine causal links that allow for the construction of heuristic beliefs and theories, offering reassuring, "ready-made" explanations. All of these factors are heightened by stress and emotions.

A Rabbit in the Clouds

Sonia Lupien, professor at the Faculty of Medicine of the University of Montreal and director of the Centre for Studies on Human Stress at the Montreal Mental Health University Institute, pointed out that every stressful situation contains one or more of four factors known by the acronym C.I.N.É.: C for the degree of Controllability of events, I for Unpredictability, N for Novelty, and É for threats to the Ego. An epidemic, a war, a natural disaster, or an economic crisis places people in a situation that is totally new (Novel), unforeseeable (Unpredictable), uncontrollable (out of their Control), and a threat to their life or identity (Ego). Combined, these factors lead to significant secretion of stress hormones such as adrenaline and cortisol. Since the work of Hans Selye, thousands of studies have demonstrated the harmful effects of stress. Nafissa Ismail, professor at the School of Psychology at the University of Ottawa and holder of the Research Chair on Stress and Mental Health, reminds us that a high level of cortisol increases the risk of developing anxiety and depressive disorders and affects the functioning of several brain regions, including the frontal cortex, by slowing its activity and impairing judgment, decision-making, problem-solving, concentration, and planning.

*****

Even if beliefs change nothing in the actual facts and have no power over the situation, they can nonetheless temporarily help to reduce anxiety and satisfy certain narcissistic traits in those who believe they possess secret knowledge unknown to the majority, feeling driven by an urgent mission to reveal it to as many people as possible. It only takes these beliefs and theories being shared by a group of initiates engaged in a similar process of denunciation, revelation of truth, revolution, and correction of wrongdoing for them to become institutionalized within protest, conspiratorial, ideological, or religious movements. This is likely why Tedros Ghebreyesus, Director-General of the World Health Organization (WHO), used the term infodemic to emphasize how "false news spreads faster and more easily than the virus, and it is just as dangerous."

PART THREE
ADDICTION TO BELIEFS AND THEIR POTENTIAL DEVIATIONS

CHAPTER 7
THE OPIUM OF THE PEOPLE

In 1844, Karl Marx stated that "religion is the opium of the people." According to him, the roots of religious belief lie in the suffering produced by oppression and the miserable living conditions of the population. Like opium, one of whose effects is to dull pain, religion places people in "the illusion that their condition is not so terrible," relying on religious moral principles that can go as far as making them believe that suffering, poverty, and misery would open to the afflicted the doors of an… artificial paradise.

Is Marx's metaphor exaggerated? Perhaps not, and that is what we will examine in this chapter. It is well known that opium, like most drugs, can create dependencies and enslave its "consumers." Beliefs can do the same for many faithful adherents. Like drug cartels, whose main objective is to promote and control trafficking operations and target vulnerable people, religions seek to recruit and convince people, sometimes, as history has repeatedly shown us, even to the point of forcing heretics to choose between conversion or punishment, paradise or hell, life or death. "Holy wars," religious crusades, jihadism, like the cruel wars between cartels and criminal traffickers, have caused and continue to cause countless victims. And what can be said of preachers, caliphs, and "gurus" who, like unscrupulous mob bosses, tycoons, and pimps, seek only to profit from humans by targeting their shortcomings, needs, and vulnerabilities.

A Rabbit in the Clouds

Religious beliefs and dogmas are not the only ones that target human weaknesses. The list of domains is enormous, think of partisanship, patriotism, chauvinism, extremism, or fanaticism of all kinds.

Is it exaggerated to put beliefs and drugs on the same level? That is what we will examine in this chapter, with reference to neuroscientific data.

Neuroscientific considerations of the need to believe

According to Michael Inzlicht, Alexa Tullett, and Marie Good of the Department of Psychology at the University of Toronto in Canada, religious beliefs fulfill one of our most fundamental needs, to create and maintain a sense of meaning in our existence and in the uncontrollable and mysterious world in which we live. Beliefs provide the illusion that the world is orderly, controlled, and understandable. However, when this coherence is disrupted, questioned, or challenged, the sense of meaning is threatened, and the believer can become destabilized, disturbed, and anxious. Inzlicht's team wanted to determine whether the effects produced by religious beliefs could be measured at the level of the brain. The region they focused on is the anterior cingulate cortex (ACC), one of whose functions is regulating cognitive conflicts, ambivalences, and uncertainties, and which also plays a role in decision-making. When you hesitate between choices, your ACC activates, the greater the dilemma, the stronger the activation.

A Rabbit in the Clouds

The researchers brought together students from various ethnic and religious backgrounds, including Christians, Muslims, Hindus, Buddhists, Sikhs, and atheists. The participants underwent the Stroop test, a task widely used in neuropsychology to evaluate certain cognitive abilities, such as selective attention and the inhibition of automatic and spontaneous responses. Numerous studies have shown that the cognitive conflicts produced by the Stroop effect are associated with relatively strong activation of the anterior cingulate cortex (ACC). The value of this task is that the subject quickly becomes aware of their errors, and those errors are unambiguous.

During the experiment, the subjects' brain electrical activity was recorded using 32 electrodes connected to an electroencephalogram (EEG). The Toronto researchers tested the hypothesis that religious beliefs could influence the level of ACC activity when subjects were exposed to cognitive conflicts. They also examined how participants' brains reacted when they made mistakes and realized it. This brain reaction, visible on an EEG trace, is called "error-related negativity" (ERN). To better understand ERN, we can think of the polygraph, or lie detector: when an individual lies, typical physiological reactions appear on the polygraph's trace.

According to Inzlicht et al., "If religion is a system of meanings that people are motivated to adopt because of the order, control, and explanations it provides, it should reduce participants' discomfort in the face of uncertainty, and this effect should be identifiable at the level of the anterior cingulate cortex (ACC) and reflected in the error-related negativity (ERN) trace."

Two groups were formed. One consisted of "ardent believers," selected based on their demonstrated commitment to their convictions when asked to rate their agreement with statements such as "In my heart, I believe my religious beliefs are truer and better than those of others," or "If necessary, I would support a war to defend my religious beliefs."

The second group consisted of "moderate believers." As the name suggests, they showed no form of fanaticism.

The results confirmed their hypothesis: ardent believers had significantly lower ACC activity and lower error-related negativity (ERN) than moderate believers. According to the researchers, the reduced ACC and ERN activity among ardent believers suggests the existence of a neurological substrate linked to religiosity and its antagonistic effects on certain brain responses. By fulfilling their quest for meaning, the intense religiosity of ardent believers managed to "numb" the psychological discomfort caused by the cognitive conflicts of the Stroop test, and very likely those provoked by unknown, threatening, and anxiety-inducing life events. The similarity between the analgesic effects of opium and those of religiosity also allows us to glimpse the possible addictive effects of the latter.

A study conducted at the University of Utah demonstrated, using functional magnetic resonance imaging (fMRI) on 19 Mormons, that the spiritual feelings of euphoria, inner well-being, and "universal love" reported by the faithful were associated with activation of the brain's reward circuits, located in a structure at the base of the brain called the nucleus accumbens. Activation of the nucleus accumbens, preceded by one to three seconds, the subjects' reports of peak spiritual feelings. When participants said they felt this spiritual well-being, their hearts beat faster, and their breathing deepened.

Such peripheral physiological changes (increased heart and respiratory rate, sweating, clammy hands, etc.) were also observed by two Swiss researchers, Yoshija Walter and Andreas Altorfer. They measured the electrodermal activity (EDA) of 37 evangelical participants who were asked to "feel the presence of God" under various conditions (with and without music, with religious and nonreligious songs, and in a resting state). The results showed that the level of electrodermal activity was associated with the intensity of the religious experience, defined by participants as "the sensation of the divine presence." Their results demonstrated that spiritual, religious, or transcendent experiences were associated with variations in the autonomic nervous system (sympathetic and parasympathetic), producing physiological reactions in peripheral organs such as the heart, respiration, and sweat glands. Similar results have been obtained during the consumption of opioids and cocaine.

These studies show that intense religious and spiritual feelings are similar in many ways to those produced by certain drugs and exogenous substances. This means that "abstract ideas and concepts," such as religious beliefs, can inhibit circuits associated with pain and discomfort and activate those linked to euphoric and pleasurable sensations. Research has also shown that meaning itself has "salutogenic" effects. For example, studies have shown that cancer patients who feel a strong sense of meaning have a greater ability to tolerate bodily illness and pain than those who find no meaning in life. Those who, despite pain and fatigue, find meaning have a better quality of life than those who have little.

These findings also suggest that the doctrinal concepts inherent in beliefs can lead to and motivate the adoption of addictive behaviors, just as drugs do. These addictions rely on the same physiological mechanisms.

Neuroscientific model of religious and spiritual addiction and obsession

For Cheryl Zerbe Taylor of Christopher Newport University in Virginia, the connection between opium dependence and addiction to religious beliefs is entirely plausible. First, the initial link identified by Taylor lies in the fact that opiates (opium, morphine, codeine), natural derivatives of the poppy, and opioids, which are semi-synthetic (heroin, buprenorphine) or synthetic (fentanyl) compounds, all possess analgesic properties. These substances soothe not only physical pain but also certain states of stress and psychological distress. These effects and sensations are achieved thanks to the neurochemical properties of narcotics (opiates and opioids). The molecular structure of opiates/opioids is similar to that of endorphins, a chemical naturally produced in the human brain, which allows them to bind to the same endorphin receptor sites (μ-receptors) in the brain and produce similar, often much stronger, effects. In addition to their analgesic effects when the body is under stress or emotional or physical pain, Taylor reminds us that the significant production of endorphins during intense activities such as sports, sex, gambling, or consumption also produces euphoric effects.

Earlier in this chapter, we mentioned the work of Ferguson et al., which showed that spiritual feelings could activate the brain's reward circuits. Two chemical substances produced by our brains were identified in these conditions: endorphins and dopamine. The consumption of opiates, like the natural release of endorphins in the brain, stimulates the production of dopamine, the neurotransmitter of pleasure, which in turn activates the nucleus accumbens and generates very powerful euphoric effects and a strong desire to reproduce them again.

In summary, addictive behaviors, whether related to drugs or beliefs, operate on two levels: on one hand, by numbing discomfort, pain, and physical or psychological distress; and on the other, by provoking ecstatic states that can lead to dissociation and loss of contact with reality. The line between hedonistic and addictive behaviors is very thin. In fact, we have here a combination of two types of motivation:

1. the pursuit of pleasure;
2. and the inhibition or avoidance of displeasure.

A Rabbit in the Clouds
The perverse effects of the "opium of the people."

When the analgesic and euphoric effects induced by natural or artificial substances cease, a form of withdrawal begins to appear. Deprived of endorphins and dopamine, the brain of the individual feels a sometimes compulsive and intense need to regain the lost sensations of pleasure or to eliminate the discomfort, unease, and suffering of withdrawal. We find here two antagonistic processes that motivate the adoption of the same behaviors (see Chapter 4: theory of opposing-process motivation). Whether it involves taking drugs or alcohol, reuniting with a loved one, or experiencing an ecstatic state, from a psychobiological point of view, the motivations are the same:

1. to experience pleasure (happiness, joy, well-being, euphoria, ecstasy…);
2. to anesthetize displeasure (produced by withdrawal, lack, boredom, distress).

To these two basic dependencies, a third may also arise by replacing the old addictive behaviors with new ones that can, in many cases, become just as harmful and troublesome.

Substitute dependencies

When a person replaces one addiction with another, it is said that they have developed a "substitute addiction." Substitute dependencies can take different forms: gambling, betting, and gaming; physical and sports activities; technology and cyberaddiction (cell phones, Internet, social media…); consumption (shopping, compulsive buying, object collecting); gluttony; romantic conquest; sex; the pursuit of power, and, of course, religion.

Steve Sussman, Nadra Lisha, and Mark Griffiths conducted a study on eleven potential addictions: tobacco, alcohol, illicit drugs, food, gambling, the Internet, love, sex, exercise, work, and shopping. Based on a review of 83 studies, they evaluated their prevalence and co-occurrence. The authors claim, based on this research, that 47% of the American adult population could suffer from an addiction disorder over a 12-month period. In another publication, these same authors note that prayer, meditation, early romantic love, and drug addiction follow the same psychobiological mechanism, involve the same regions of the brain, produce similar emotional states, and cause similar withdrawal syndromes. As Steve Sussman writes:

"Prayer, meditation, early romantic love, and drug abuse have in common the activation of dopamine-release processes in the brain and the attainment of intense emotional states. In this sense, reliance on a higher power can function as a substitute addiction, replacing the psychobiological functions once occupied by drug use, [whose perverse effect may be] in some cases, avoiding working on one's underlying problems and relying on the Higher Power to solve them."

* * * * *

In conclusion to this chapter, here is the testimony of a former Scientology follower expressing his difficulty in breaking free from his beliefs:

A Rabbit in the Clouds

"I once dabbled in Scientology. I took distance courses to study the basic material (books and lectures). I ended these courses before completing the program. I even left angry after reading and watching documentaries on Scientology. I wrote a letter to the church in which I… let them have it, so to speak, and asked them not to contact me again.

A few months ago, I began to miss the church's theories. In fact, their teachings make sense in my head; that's the problem. I find what's explained in their books quite logical, which made me start taking distance courses again to restudy the material. What's tiring with them is that there's always someone assigned to you who texts or calls to check on your studies, pushing you to study faster so you can move on to another course, etc.

I got a call from this person yesterday. Not being someone who asserts myself easily, she convinced me to send in three lessons this week so I could start a new course this weekend… But now I realize I'm back in the same situation as before: I like the material I'm studying, but I feel pressured and have no freedom to learn as I wish.

How can I leave the movement for good and understand that it's not something good? I don't understand myself… It's expensive; I've read books that testify to people's experiences in Scientology; I've watched documentaries… I know about the abuses that happen there, physical, mental, and financial. Why am I still drawn to Scientology, and why, despite everything I know that's wrong with them, do I still want to keep studying? I shock myself… And even now, if I want to put an end to it again, how do I do it? I hate confrontation, but at the same time, ignoring the church's calls and texts will only make them keep contacting me. I feel so lost in all this."

CHAPTER 8
THE DEVOTED ACTOR

Beliefs likely appeared at the same time as human beings. Thanks to our cognitive faculties (which distinguish us from animals), including an inexhaustible imaginative power, humans have been able to invent writing, electricity, the space shuttle, modern medicine, computing, the Internet, artificial intelligence, the nuclear weapon, without forgetting great artistic, literary, and cinematic works, as well as fiction, legends, myths, superstitions, and… religions.

Beliefs have also been used to serve economic, political, and religious interests. In many countries around the world, religions have exercised immense power over individuals and societies. They have controlled education, imposed coercive and constraining moral norms and rules, and prescribed practices and rituals from birth to death. Hand in hand with politics, they have been instrumentalized for ends not always benevolent, and, sadly, far too often base, abject, and shameful. One need only recall Nazism, the Ku Klux Klan, Alcalisa, the Inquisition, the Crusades, etc.

Among the most dismaying aspects of human nature is this potential for hatred that can lead human beings to fight, kill each other, and even sacrifice their own lives in the name of ideologies and beliefs. Specialists in cognitive science and neuroscience have sought to understand political and religious violence by conducting studies with members of extremist, jihadist, patriotic nationalist, and conspiracist groups such as Canon. Their work led them to identify a particular type of individual who forms the backbone of these groups, organizations, and movements and, more than any other member, grants them boundless power. They call them "devoted actors."

SACRED VALUES: the bases of fanatical terrorism

A study conducted among Americans and Palestinians found that, in conflict situations, resources that are contested or claimed can become sacralized in the eyes of adherents (for example, lands and holy sites, temples, monuments and texts, symbols, flags, fetishes, idols…), particularly when religious, political, and military bodies seize upon them for their own ends. Perceived threats against their ingroup also lead the faithful to regard their sacred values as preferences they must protect.

The same researchers also conducted a study among a group of Palestinians in the West Bank and Gaza. They found that conflicts centered on sacred values are particularly difficult to resolve because they cannot be compensated for by fungible goods, objects, or resources. For them, sacred values are non-negotiable, and usual bargaining attempts, such as offering indemnities or material compensation, are not only ineffective but risk increasing moral outrage and spurring violent acts.

Nafees Hamid, a researcher at the International Centre for the Study of Radicalisation (ICSR) based in the Department of War Studies at King's College London, points out that despite being called "sacred," these values are not solely religious. For some, freedom of expression, sovereignty, democracy, power, or money may also be considered sacred. For others, sacred values might include devotion to a charismatic leader, the resurrection of a caliphate, the return of Christ, or the establishment of sharia as a universal rule. When these sacred values are violated, those who adhere to them, especially if they belong to a group that shares the same convictions, are often ready to do anything to defend them. Devoted actors truly begin to take shape when identity fusion and sacred values interact.

Surveys conducted among fighters against the Islamic State (IS) point in the same direction, emphasizing so-called "non-utilitarian" aspects that combatants, whether secular or religious, consider sacred or spiritual. They show that the importance attached to sacred values and identity fusion among devoted actors manifests in:

1. Commitment to non-negotiable sacred values and to the group to which they are strongly bound (identity fusion);

2. A willingness to abandon their loved ones for their cause and mission;

3. The perceived spiritual strength of the ingroup relative to enemies is more important than relative material strength.

Hammad Sheikh, Ángel Gómez, and Scott Atran reached similar conclusions. They conducted an experiment with 260 subjects from Moroccan cities and neighborhoods associated with militant jihad and 644 Spaniards, fairly representative of the country as a whole. When affiliated with a close-friends group that considered sharia sacred, the Moroccans expressed willingness to make costly sacrifices for implementing this law and were more favorable to militant jihad; Spaniards involved in groups that firmly believe democracy is sacred were likewise willing to make costly sacrifices for it, especially when told it was threatened by enemies. The researchers note that their "results paint a portrait of devoted actors who feel viscerally bound to a fundamentalist group, who hold strong values, and who are prepared to make extreme sacrifices (including willingness to fight, kill, and die) when they believe their values are threatened."

For Hammad Sheikh, "human beings are capable of believing in the absurd [...] beyond reason and evidence. This kind of devotion has led to some of our greatest achievements. Yet, as our research shows, devotion also has a dark side."

Personal spiritual convictions (distinct from religiosity) are also strongly associated with the willingness to fight at the cost of significant sacrifice for the group. This was revealed by a large study of 3,285 subjects in six countries across four continents, which corroborates the same observations. This individual factor is further exacerbated by the loyalty of the faithful to members of their group who share the same spiritual convictions.

We have seen that violent extremism is often motivated by devotion to abstract ideals such as the nation, divine law, and sacred values and beliefs, while being much less attached to material benefits. These factors also serve to define and identify one's own group membership. Moreover, behaviors aimed at denouncing, opposing, defending, or fighting what might threaten their beliefs and ideals often intensify when there is contestation, opposition, discrimination, ostracism, or social exclusion.

Neuroscientific foundations of extremism

Thanks to functional magnetic resonance imaging (fMRI), neuroscience reveals fundamental information about extremist and radical beliefs, devotion and attachment to sacred values, willingness to fight for these beliefs and values, and the propensity to adopt violent behaviors against opponents, up to risking one's own life.

A team of researchers at Emory University in Atlanta, USA, tested the integrity of and attachment to sacred values by proposing to devoted actors that they "sell their personal values" in exchange for real money if they accepted. Placed in an fMRI scanner during the negotiations and the adherents' refusal to sell their sacred values, Berns et al. observed increased activation in brain regions involved in adopting semantic rules: the left temporoparietal junction (TPJ) and the ventrolateral prefrontal cortex (VLPFC). According to them, this suggests that sacred values influence behavior and decision-making based on deontic rules, that is, norms or moral principles dictating what is considered "morally right or obligatory", rather than on a utilitarian "cost/benefit" evaluation.

Another fMRI experiment, this time conducted by Clara Pretus with 38 young Moroccans living in Barcelona who had expressed willingness to engage in violent acts associated with jihadist causes. All participants took part in an online ball-toss game with three virtual players of typically Spanish appearance. The game simply involved passing the ball among them. Two groups were formed. Half were randomly placed in a control group (called the "social inclusion" group). In their case, the game was programmed so that every time the Moroccan subject sent the ball to a Spanish player, the latter returned it afterward. The other half were placed in the "social exclusion" group. The game was the same, except that after one or two passes to the Moroccan player, the three Spanish players began to play only among themselves, thus excluding the jihadist sympathizer. Subsequently, in turn, the subjects were placed in the fMRI scanner, where they had to assess their willingness to fight and die for sacred values or for non-sacred values. Among the "social inclusion" group, strong activation was observed in brain areas associated with sacred values (the left inferior frontal gyrus). Among the "social exclusion" group, the results differed in two ways: 1) activation of the left inferior frontal gyrus was present not only for sacred values but also for non-sacred values; and 2) their explicit willingness to fight and die for these "non-sacred" values increased! In other words, these results show that non-sacred values become more like sacred values in terms of neural activity and willingness to fight and die when participants experience social exclusion.

Social exclusion could therefore predispose people to fight and even die for values that were not sacred to them initially. It was also found that following social exclusion, non-sacred values showed an increase in emotional intensity measured in the amygdala, comparable to that observed for sacred values.

A Rabbit in the Clouds

At the end of a study using fMRI with radicalized individuals, Nafees Hamid and colleagues identified neurocognitive processes specific to the commitment of devoted actors to their sacred values. They even speak of a "neural signature" specific to the willingness to fight and die for them.

More in-depth research was conducted with 30 young Pakistani men belonging to the armed radical movement Lashkar-e-Taiba, which supports Kashmiri fighters linked to Al-Qaeda. When examining their willingness to fight and die for their sacred values, the researchers observed deactivation of areas of the dorsolateral prefrontal cortex (DLPFC), associated with deliberation and self-reflection. This could mean that these executive functions are not used to decide the level of violence with which to act in the name of their sacred values. In fact, their explicit willingness to fight and die for them rested on brain activity in the ventromedial prefrontal cortex (VMPFC), known for evaluating subjective values and social rewards.

Interesting facts: when these young devoted actors (the young radical Pakistanis) were told that their less radical fellow citizens disapproved of their violent behavior to achieve their ends, they reduced their commitment to resorting to violence. This decrease in their commitment to fight furiously coincides with reactivation of brain areas involved in deliberation and self-reflection (the DLPFC). This suggests that the lack of approval from their more moderate fellow citizens influenced these young radicals to reflect (thereby reactivating their DLPFC) rather than react with unconditional automaticity to their sacred values.

These data suggest that this differential brain activity could be part of the neural signature of the devoted actor when he expresses his determination to sacrifice himself to defend and uphold his sacred values, but not for non-sacred values. This signature would be characterized by hypofunctionality of regions involved in higher executive functions of the brain (including the DLPFC) and overactivity of more primary areas of the VMPFC. It should be noted that the emotional states produced by the VMPFC can significantly influence reasoning and decision-making processes. Consequently, rapid reflection regulated by sacred values prevails over self-deliberation when making high-stakes decisions within extremist groups.

As Nafees Hamid points out, argumentation and debates that seek to challenge and oppose the beliefs and sacred values of devoted actors are often fruitless and can even provoke the opposite of the intended effect, by triggering a sense of exclusion among those who hold them.

Elizabeth L. Paluck, professor of social psychology at Princeton University, led field research in Central Africa and the United States. Among these were two one-year field experiments conducted in Africa: the first in Rwanda and the second in the Democratic Republic of the Congo (DRC).

A Rabbit in the Clouds

Paluck spent a year in Rwanda, where she assessed the limits and influence of the media on shaping prejudices, transmitting norms, and changing behavior. In several regions chosen at random, they broadcast a fictional radio soap opera on the theme of reconciliation and reducing violence in two Rwandan communities. They later assessed and found that citizens' personal convictions regarding reconciliation or cooperation had not changed. This means that Hutus and Tutsis who had not supported reconciliation a year earlier still did not approve of it after the soap opera. What did change was their perception of what the other group might believe: Hutus now perceived that Tutsis believed in reconciliation and cooperation, and vice versa.

In another experiment conducted in the Congo, Elizabeth Paluck examined whether mass media that offered listeners the opportunity to express themselves and discuss could help reduce intergroup conflict. In a randomly selected portion of the population in eastern DRC, a talk show was broadcast that encouraged listeners to hear the opinions and perspectives of opposing groups, to air their own grievances and viewpoints with those groups, and to debate with them. In another region, they broadcast only a fictional soap opera, as in Rwanda. Compared to individuals exposed only to the soap opera, listeners and participants in the talk shows were more intolerant, more focused on their grievances, and less inclined to help members of the opposing community. In all cases, the radio program hardly changed listeners' personal convictions.

A Rabbit in the Clouds

For Paluck, these results underscore the limits of well-intentioned media that provide platforms for discussion on political issues, which could worsen the situation because of the controversial nature of the dialogues. These findings should also prompt reflection on the relevance of contentious debates, regularly found on CNN, BBC, Fox News, TF1, etc., in order to carefully assess their potential to harm the proper functioning of a deliberative democracy.

Based on the research presented in this chapter, two approaches are proposed to change or prevent extremist behavior. The first is to steer devoted actors' perceptions toward common social norms that value socially beneficial behaviors, instead of prompting them to discuss, argue, and debate everyone's beliefs and values.

The second approach is to rely on moderate and influential peers who oppose the excessive, radical, and violent behaviors of devoted actors, in order to push them toward self-reflection (thus reactivating their DLPFC) and to consider options other than violence.

Research has also shown the benefits of introspection and deep personal work. For example, a 2018 systematic review showed that the long-term changes from mindfulness-based interventions were primarily associated with increased activity in the insular cortex. Numerous fMRI studies have demonstrated the role of the insula in social cognition and empathy. Empathy is defined as "an affective and cognitive response toward others, involving understanding and experiencing another's feelings as well as the ability to adopt their subjective point of view. Several processes underlie this response, including interoception, self-awareness, and social cognition, which depend on several cortical and subcortical regions, including the insula."

* * * *

I have mentioned several times that despite the right and freedom to believe, we should in no way underestimate the role of beliefs in the daily lives of individuals, social groups, and societies. Although we have emphasized the perverse and disastrous effects of devoted actors, combined with sacred values and identity fusion with their ingroup, we must also acknowledge their contribution to just and beneficial causes. We all benefit today from the devotion of actors who worked peacefully in the past for racial and gender equality, minority rights, and individual freedoms. Today, devoted actors are engaged in ecological and environmental causes seeking ways to correct the course of a global climate crisis; to resolve and prevent deadly wars and conflicts; to implement social interventions aimed at preventing juvenile delinquency and the devotion of young people to criminalized gangs…

The reader will have understood that, without ignoring the invaluable contributions of Martin Luther King, Nelson Mandela, Mahatma Gandhi, Emmeline Pankhurst, and many others, there are also actors whose lack of humanity, altruism, and compassion reveals the dark side of human nature.

If, from an individual standpoint, irrational beliefs can be at the origin of certain behavioral, affective, and emotional disorders, when they add up and are inscribed within a collective belief system, practices can take bizarre, eccentric, or far-fetched forms and even drift into trivial rituals such as genital mutilation, scarification, and extreme savageries by fanatical groups. It would also be a mistake to think that violent and extreme acts are only the work of unhinged, solitary psychopaths. That does exist, of course, but we should be aware and vigilant that there are also adherents fused within groups who are ready to do anything for a cause they believe to be sacred.

We have all been marked by the September 11 attacks and those on Charlie Hebdo and the Bataclan, committed by extremists. But there have also been assaults committed by "devoted patriots," such as during the Capitol insurrection in Washington on January 6, 2021, and the attempted coup on January 9, 2023, in Brasília, when the National Congress building, the presidential Planalto Palace, and the Supreme Federal Court were invaded and vandalized by supporters of former President Jair Bolsonaro. Upstream of these violent acts are individuals convinced they are acting for noble causes and values such as democracy, law, justice, and freedom… Among Donald Trump's supporters are devoted actors whom he himself feeds with the same ingredients as extremist groups. It took only a spark to ignite the powder among the most fervent fanatics who descended on the Capitol in January 2021, with calls such as: "You will never take back our country with weakness. You have to show strength, and you have to be strong…" A few hours after the insurrection at the Capitol, Trump persisted and tweeted: "These are the things and events that happen when a sacred and landslide election victory is so viciously and unceremoniously stripped away from great patriots who have been badly and unfairly treated for so long." In this comment, we find a sampling of key ingredients:

, the violation of a sacred value: democracy;

, the vicious stripping away of an electoral victory;

, the identity fusion of devoted actors, the "patriots," with their country.

This chapter aims to show that beliefs, often trivialized and even ridiculed, can have "extreme" consequences at many levels. Let us not forget that upstream of individual behaviors such as addiction to gambling, superstitions, prayer, crime, and violence, there are deeply rooted beliefs. The same goes for conflicts, wars, genocides, etc. As was the case with Napoleons, Hitlers, and Stalins, so it is today with unscrupulous political leaders. The few examples drawn from scientific research presented in this third part help us understand how far convictions can go. This includes that former U.S. president who claims that, because of his fame and fortune, he can "grab women by the pussy" and then do what he wants; those heads of state who declare a bloody war under the spurious pretext that they want to "carry out an operation against Nazism or to destroy Hamas"; who claim it is justified to sacrifice civilian populations, and who oppose the sovereignty of the Palestinian people in Gaza; who assert that Israel must retain "security control" of the territory; or a dictator who declares that his country "will not hesitate to retaliate with nuclear weapons if it is itself 'provoked' by nuclear arms." Moreover, it is chilling to know that they have at their service devoted actors who are submissive and ready to do anything to defend their sacred homeland.

Our society, supported by the Human Sciences, faces an immense challenge to correct these deviations and prevent all their disastrous consequences.

PART FOUR
SCIENTIFIC THINKING AND CULTURE AS AN ANTIDOTE TO IRRATIONAL BELIEFS

PREAMBLE

The insatiable and inexhaustible need to know and explain the world...

One of the great particularities of human beings lies in the fact that they question themselves and feel the need to find answers to their questions. This need drives them not only to "know," but also to "understand" and to "explain" the world around them and the universe in which they exist. The cognitive abilities that serve this need are essentially the same as those that make humans imagine causal links where none exist. This insatiable quest to know at all costs the causes and effects of almost everything likely finds its roots in the fundamental needs of individual and collective survival. Our history shows that the survival of individuals, their offspring, and their communities has essentially rested on a succession of trials and errors. When deprived of science and technology, they use whatever they can to survive, driven, despite everything, by the desire to "know better in order to act better."

Among all that has evolved in human history, there is, of course, the methods of acquiring and transmitting an ever-expanding body of knowledge. The creation and dissemination of beliefs have also benefited from certain technical advantages, most notably the Internet and social media. Many beliefs have been structured through systems that were built around them. Religious and dogmatic institutions are good examples. The rigidity of such systems of belief is so great that many dissenters, rebels, and heretics have had to answer before inquisitorial tribunals and pay with their freedom and even their lives.

Moreover, it may be that uncertainty, doubt, challenge, and counter-evidence have stimulated another need, the need to discern truth from falsehood.

Rationally, we easily admit that it is utopian and illusory to hope to protect ourselves entirely from all the biases that contribute to the formation and adherence to various beliefs and theories. Yet, that does not prevent us from still seeing, even today, obsessive and sometimes compulsive behaviors such as: drawing one's opinions from self-proclaimed influencers and bloggers; taking one's child to the local quack; consulting fortune tellers or reading one's horoscope in the daily paper…

Alongside all the irrationality of which humans are capable, they dream of finding an infallible path to truth. Undoubtedly, our society would benefit from it at many levels (legal, medical, political, scientific, religious…), but things are not that simple.

The question remains: what means or processes other than blind trial and error can we use? Since there is no absolute method, perhaps we should look at what has helped us most to understand our world, its organization, its functioning, and to advance as a society, to survive natural disasters, to defeat many diseases, to extend our life expectancy, to adapt, and to unlock some of the universe's great mysteries: science.

A Rabbit in the Clouds

Science is not a belief, nor does it function like religions or dogmatic institutions frozen in time. It proceeds methodically, using approved instruments and techniques, surrounding itself with multiple safeguards, engaging in self-criticism and self-correction, publicly exposing its discoveries to meticulous peer examination, and constantly calling for those discoveries to be tested and reproduced.

CHAPTER 9
THE MEANS OF ACQUIRING
KNOWLEDGE

Although the scientific method is not the only means of acquiring knowledge, it remains by far the most effective process. Epistemologists generally distinguish six different ways of acquiring knowledge: superstition, intuition, authority, rationalism, empirical observation, and the scientific method. Let's look at them one by one.

The Five Non-Scientific or Pseudoscientific Approaches to Acquiring Knowledge

1. Superstition

Superstitions are undoubtedly among the oldest and most persistent ways of explaining things. A superstition can be defined as a belief to which one adheres and reacts, attributing to it causes and powers that are in fact unlimited. During the Middle Ages, for example, "the plague re-emerged shortly after the appearance of a comet. Pope Callixtus III ordered that all church bells be rung every day at noon to ward off this curse. That is, incidentally, where the custom of the Angelus originated."

Superstitions exist on nearly every subject imaginable. They are seen daily in casinos around the world: superstitious gamblers associate rituals, gestures, or thoughts with winnings and start repeating them. Similar ritualistic behaviors are found among ordinary people and even professional athletes, serious businesspeople, and politicians.

In 1948, psychologist B. F. Skinner demonstrated that superstitions are learned behaviors. His experimental subjects were pigeons. He placed a pigeon in a cage and, purely at random, dropped food into its tray. Curiously, the pigeon began making absurd associations between its actions and the arrival of food. Gradually, it developed strange ritualized behaviors, turning in circles, jumping, or cooing, that in fact had no effect on whether food appeared. Since the reward sometimes arrived right after a random movement, the pigeon associated its act with the outcome and repeated it. Skinner explained that "the bird behaves as if there were a causal relationship between its behavior and the presentation of food, even though no such relationship exists."

Across cultures, tribes, and societies, superstitions take ever more extravagant and unusual forms. Even today, entire industries thrive on them, such as faith healers, casinos, lotteries, sports betting, and more.

2. Intuition

Intuition is the second "non-scientific" approach to knowledge. It refers to a way of knowing that is not based on reasoning or inference. Intuitive thought arises spontaneously and can be useful when quick decisions must be made without analyzing details, sometimes a valuable skill for adaptation or survival, or when one has deep expertise in a field. Decisions are made swiftly, without thorough reflection or objective analysis. While intuition can occasionally yield the right result, countless situations show how easily it can mislead and produce erroneous conclusions.

3. Authority

This form of knowledge acquisition relies on accepting information or explanations issued by a recognized or respected authority. The argument from authority rests essentially on the credibility granted to an individual, without real assurance of the value or validity of their claim or of the method by which they reached it.

Ordinary people frequently appeal to authority when seeking information, making decisions, or assessing situations, something often justified, as when consulting a specialist or an expert's opinion. But history reminds us of the abuses of authority by religions over millennia. Built upon the words of a higher power and founded on irrevocable authority, followers of all faiths were required to submit and defer to it to know the "truth."

Today, information and misinformation circulate faster and reach farther than ever. Authority now extends well beyond kings, chiefs, or preachers, it permeates every domain, from pseudo-experts to miracle-cure charlatans and online scammers.

A striking example is Dr. Luc Montagnier, Nobel Prize winner in Medicine (2008) for discovering HIV in 1983. Celebrated for that achievement, he later strayed into promoting partisan, unscientific positions, claiming, for instance, that COVID-19 variants were caused by vaccines. The scientific community harshly criticized him for promoting dangerous pseudoscience. Yet many internet users echoed his statements to support their beliefs, citing his title as proof: "He's a Nobel Prize–winning scientist, Dr. Montagnier said it!" This is a classic illustration of the authority bias described in cognitive psychology.

4. Rationalism

Many might assume that logical reasoning is more reliable than superstition or intuition. Because it relies on reason and logic, one might think that following the right reasoning process ensures a correct result. Not necessarily, the matter is not so simple.

A common syllogism shows the limits of rationalism:

- Premise 1: If it rains, the game is canceled.
- Premise 2: It is raining today.
- Conclusion: Therefore, today's game is canceled.

If we accept both premises as true, few would doubt the conclusion.

But reverse the order slightly:

- Premise 1: If it rains, the game is canceled.
- Premise 2: Today's game has been canceled.
- Conclusion: Therefore, it is raining today.

Although both premises are valid, the conclusion is not; many other factors could explain the cancellation (illness, forfeiture, equipment failure, lack of players, etc.).

Thus, rationalism is useful but limited. In many cases, we cannot account for every variable needed to reach precise and valid conclusions. Even a logical reasoning process built on sound premises does not guarantee a true conclusion.

5. Empiricism

Empiricism "holds that knowledge is founded on the accumulation of observations and facts from which laws can be derived." This approach asserts that people derive knowledge from their perceptions and experiences, often summed up by "I saw it or lived it, so it must be true."

Although not scientific in itself, empiricism often stimulates scientific inquiry. In many fields, empirical observations have raised questions that led to more rigorous research, revealing causal links and valuable explanations. In that sense, science is empirical, since it relies on structured, methodical observation, but an empirical approach is not necessarily scientific.

Rationalism, empiricism, and even intuition can each play valuable roles in the scientific process.

In short, in the perpetual search for causality, intuitions, observations, and logical reasoning all face a virtually incalculable number of factors that can influence phenomena in varying degrees. It is therefore hardly surprising that we often take wrong turns and make mistakes.

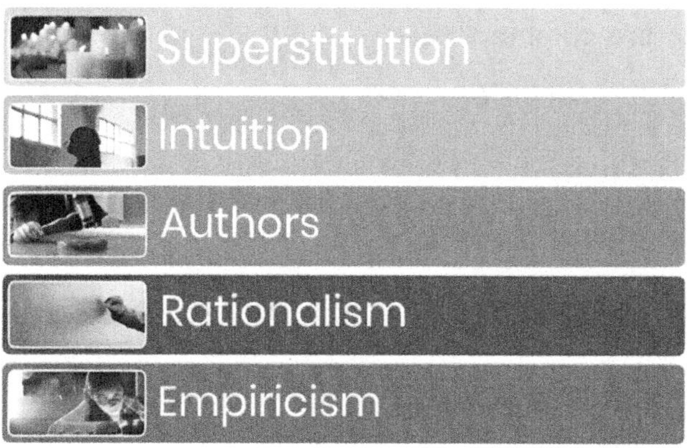

Figure 1. The 5 non-scientific or pseudo-scientific approaches to knowledge acquisition

The Prosecutor's Fallacy

The prosecutor's fallacy takes its name from the legal field. It occurs when one equates the probability of acquitting a defendant against whom there is strong evidence with the probability of convicting an innocent person based on similar evidence. In both cases, we commit what is known as the prosecutor's fallacy.

- Convicting an innocent person corresponds to a false positive.
- Acquitting a guilty person corresponds to a false negative.

That is why, in law, there exists a moral principle stating that "it is better to acquit several guilty persons than to condemn a single innocent one." In other words, from a legal standpoint, a false negative is preferable to a false positive.

However, the prosecutor's fallacy is not limited to the judicial world. It also occurs in many other contexts, revealing the frequent flaws of intuition, empirical observation, and logical reasoning. Even when these are combined, as we shall see, nothing guarantees the accuracy of the conclusions we draw.

Example: A Medical Test

Imagine you decide to go to a screening center to be tested for virus X. The test you take is said to be 99% accurate.

Your test result comes back positive. Naturally, your immediate question would be:

"What is the probability that I am actually infected with virus X?"

And the spontaneous answer that may come to mind is:

"My test is positive, so I'm almost certainly infected with virus X."

Your reasoning might sound logical:

- "The test I took is 99% reliable."
- "My result is positive."
- "Therefore, there is a 99% chance that I am infected with virus X."

But is that reasoning correct?

No! Your probability of actually being infected with the virus is not 99%. You've just committed the prosecutor's fallacy. Here's why:

Understanding What "99% Accurate" Really Means

This figure comes from experimental testing showing that out of 100 people actually infected with the virus, the test correctly detects 99 of them and misses 1 (a false negative).

A Rabbit in the Clouds

To determine your real risk of being infected with virus X, we need at least three variables:

1. The test correctly identifies 99% of infected individuals (1% false negatives).

2. The test gives false positives 2% of the time, meaning it wrongly declares 2 out of every 100 healthy people as positive.

3. The prevalence of virus X in your region, that is, the percentage of people who are truly infected (whether they know it or not). Let's assume the prevalence is 1% (1 person out of 100 carries the virus).

Let's Do the Math

Suppose we randomly select 10,000 people from your region to be tested for virus X.

- Prevalence: 1% → 100 people are actually infected.
- The test correctly detects 99% of them → 99 true positives and 1 false negative.

That leaves 9,900 healthy people. Among these:

- 98% will be correctly identified as negative → 9,702 true negatives.
- 2% will test falsely positive → 198 false positives.

So, across all 10,000 tests, the results are:

- 99 true positives (infected people correctly identified)

- 198 false positives (healthy people incorrectly identified as infected)

That's 297 positive test results in total.

Now, to find the probability that a person who tested positive is actually infected:

99/297 = 0.333……

In other words, given the accuracy of the test, its false-positive rate, and the local prevalence, the probability that you are truly infected is about 33%, not 99%.

Figure 2 below illustrates the calculation process.

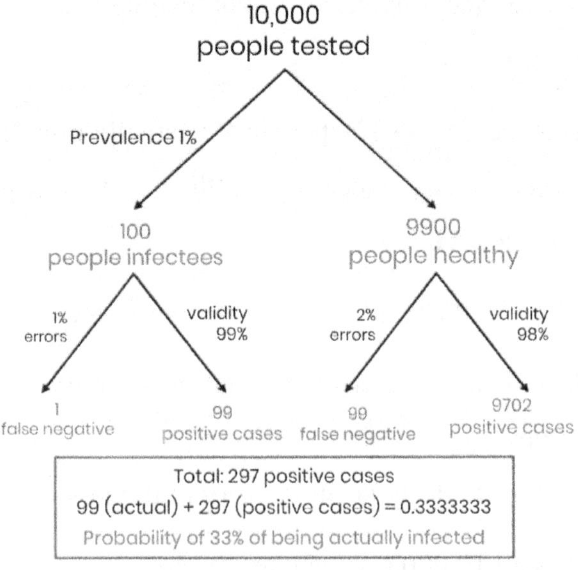

Figure 2. Conditional probability calculations in the case of a medical test.

A Rabbit in the Clouds

A fundamental point emerges from this simple demonstration: merely adding, removing, or changing a single parameter in the data can lead to a completely different result. For instance, if the prevalence of virus X infection in the population dropped from 1 in 100 to 1 in 1,000, then only about 10 people out of 10,000 would actually be infected at the time of testing. The calculations would then show a probability of being truly infected of less than 4%.

Now imagine adding more variables to the equation, the type of symptoms, comorbidities (e.g., diabetes, hypertension, obesity), whether someone lives in an urban or rural area, the varying conditions under which the tests are analyzed (differences in lab procedures, human error rates, hygiene conditions, etc.). In an ideal world, each of these factors would need to be weighted, included in the equation, and their influence measured.

This is what is known as conditional probability.

A misunderstanding of this concept can easily lead to statements such as:

- "Tests are useless since a positive result on a test that's 99% accurate means there's only a 33% chance of being infected."
- "How can a test that's 99% reliable give such poor results?"
- "This proves that you can make numbers say whatever you want."

Let's focus on this last comment, "you can make numbers say whatever you want."

A Rabbit in the Clouds

That belief is false. Numbers do not "say whatever they want"; quite the opposite! When data are properly collected and calculations are correctly carried out, numbers are highly explicit. They tell us exactly what they refer to (objects, healthy or sick individuals, addicts, rich or poor, etc.); they specify their unit of measure (count, percentage, proportion, etc.); and they have meaning that depends on the field of expertise in which they are used (medicine, engineering, accounting, and so on).

It is therefore not numbers that mislead us, but people, those who throw figures around carelessly, without specifying what they actually refer to. The same applies to words. From a literary standpoint, words precisely define an object, idea, or quality. That doesn't prevent people from using them carelessly or irreverently. Words remain exact; it's the user who isn't. Unfortunately, this bias is widespread!

That is why, and we will return to this later, it is essential to verify where information comes from and what its sources are; to understand the methods used for data collection; which instruments were used to obtain the measurements; and what statistical analyses were performed. Above all, we must ensure that the results are properly interpreted and accurately reported.

CHAPTER 10
DISTINGUISHING WHAT IS SCIENTIFIC FROM WHAT IS NOT

If, in acquiring our knowledge, we rely on the non-scientific approaches described in the previous chapter, we leave ourselves at the mercy of all kinds of simplistic, fanciful, and irrational beliefs and theories. That is why the scientific approach and method remain the most reliable means we have. However, this approach must meet very strict and rigorous conditions; otherwise, it risks sliding into pseudoscientific beliefs. Here are the four main conditions:

1. In science, we must be able to observe the phenomena being studied (directly or indirectly). Let us remember that science is empirical, meaning that it uses observation and experimentation to acquire knowledge. To be able to conduct a scientific study, we must be able to observe the objects, phenomena, or subjects being studied. Many phenomena can be directly observed. If not, we must use instruments to do so. We can thus observe many manifestations or traces indirectly. For example, in psychology, it is generally possible to observe certain behaviors directly (speaking, walking, shouting, crying, etc.). However, there are several mental processes that cannot be directly observed (a thought, a feeling, a sensation, a state of consciousness, etc.), but which can be observed indirectly. Researchers may ask the subject to name what they are thinking

or feeling; they may also use instruments to take physiological measurements (electrocardiogram, polygraph, EEG, MRI, etc.).

The examples are numerous, and the means of observing and measuring are just as varied.

However, when a supposed phenomenon is neither observable nor measurable, such as the existence of a spiritual entity (God, the devil, the soul, etc.) or a mysterious force (diabolical, supernatural, metaphysical, etc.), we are dealing with manufactured beliefs, nothing more. Some readers may say that just because we have not proven the existence of a phenomenon does not mean it does not exist or that we will never be able to prove it. That is possible. But until then, we are navigating the realm of belief.

2. We must also be able to measure the phenomena observed (directly or indirectly). We can measure frequency, rate, percentage, degree, number, strength, etc. As with observation, measurement can be direct or indirect (that is, by using tests or measuring instruments). Measuring is therefore fundamental in science because it makes it possible to quantify observations and eventually carry out more in-depth statistical analyses on them.

From a scientific standpoint, making a factor measurable is called operationalization. For example, we may want to verify whether motivation is a factor in athletic success. We must therefore define, in operational terms, what motivation is; otherwise, it is impossible to study it. Is it the number of hours spent training, their frequency, or something else? In addition, we must also operationalize the variable "athletic success": is it the number of points scored, the number of victories, or the number of competitions? Only under these conditions can researchers observe and measure the relationship between these variables.

3. We must also be able to reproduce the observations and measurements. Reproducibility is fundamental in science. In principle, if a phenomenon truly exists and has been objectively observed and measured, we should be able to reproduce the observations and measurements, thus ensuring that we are not dealing with an epiphenomenon, an anecdote, an isolated incident, or pure fabrication.

4. The fourth condition is refutability. This is an essential criterion in science. As philosopher and epistemologist Karl Popper stated: "A theory is scientifically acceptable only if it is 'refutable', meaning that it can be subjected to experimental tests to verify the agreement between its theoretical predictions and the observations. A hypothesis that could not be refuted by any experiment or observation is not scientific."

Popper gave an excellent example of this condition when he pointed out that the concept of God, like many religious, esoteric, superstitious beliefs and myths, as well as well-known theories such as psychoanalysis or Marxism, fall outside the realm of science, because their existence cannot be refuted and therefore cannot be confirmed either. This is the case for a countless number of concepts, beliefs, and theories of all kinds.

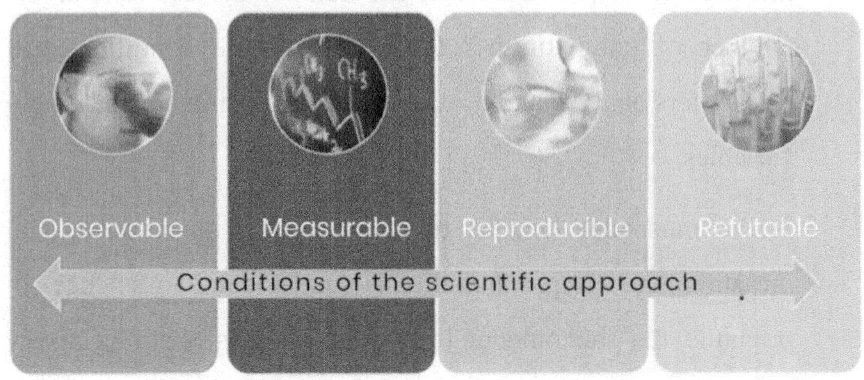

Figure. Conditions of the scientific approach.

Acquiring Knowledge Through the Scientific Method: The Stages of a Scientific Approach

We saw, with the example of virus X, that a seemingly insignificant detail can drastically change the interpretation and conclusions of our observations. The vast majority of natural and human phenomena we study are so complex that it is impossible to include every factor in our approach.

In math class, we learned to solve equations with two or three unknowns. But in reality, we constantly face equations with multiple unknowns. In our quest for knowledge and explanations, we often give in to the temptation to exclude or ignore numerous factors in favor of others, usually simpler ones, that align with our own beliefs or seem more valid, and we end up taking shortcuts. The questions we ask ourselves are no less relevant, but before claiming to know the real answers, we must take the time to consider a wide range of data and information.

Analogy Between a Criminal Investigation and the Scientific Approach

To better grasp what the scientific method represents, let's refer to a criminal investigation.

In a criminal investigation, establishing evidence is essential to determine whether a suspect is guilty "beyond a reasonable doubt." And that's no small feat! Criminologist Kim Rossmo of Texas State University describes three operational phases of an investigation:

– Phase 1 consists of collecting, evaluating, and analyzing evidence. The evidence must meet two very strict criteria: relevance and reliability. "Relevance concerns the 'weight' of a piece of evidence, meaning the strength with which it supports or refutes the guilt of a suspect, or the importance it lends to a given theory about the crime under investigation…" Reliability refers to the precision or accuracy of the evidence. Even if a piece of evidence seems relevant, if it is flawed, it has little or no probative value. For example, if a DNA sample found at the crime scene matches a suspect's DNA, that is relevant evidence. However, its reliability could be questioned if the suspect regularly visits the location, making it normal for his DNA to be found there.

– Phase 2 involves identifying, ranking, and evaluating suspects. This process of identifying an offender relies on evidence, whether it comes from the crime scene (e.g., fingerprints, hair, clothing) or from focusing on a particular suspect (e.g., criminal record, eyewitness testimony).

– Phase 3 aims to establish the connections between the evidence gathered and the most likely suspects. It is this third phase that leads to conviction or acquittal.

These three phases highlight the crucial difference between a rigorous investigation that can lead to a conviction and mere presumptions or allegations. Of course, even such a process is not infallible; judicial errors exist.

The Scientific Approach

The scientific method is defined as a set of procedures and techniques that allow us to verify a previously formulated hypothesis. It is the method used to answer questions that distinguishes "science" from "non-science."

The analogy between the judicial and scientific domains illustrates that both rely on a process of confronting hypotheses with reality, while striving to remain as objective and impartial as possible.

We all know that nothing in this world is perfect. But when the process is properly carried out, the risk of error is minimized. Can there still be biased judgments, poorly conducted, or even fraudulent research? Unfortunately, yes. That is precisely why "safeguards" exist, which we will examine in the next chapter.

A Rabbit in the Clouds

In law, there is the right to the presumption of innocence and the right to a full and fair defense before an independent court. Moreover, in the event of a conviction and sentencing, it is possible to appeal the judgment before another court, usually composed of several judges.

Similarly, if we carefully read a scientific report, we notice that the authors impose upon themselves a strict discipline and a duty to critically assess their own work throughout the process. They must learn to step back and examine their study as if it had been conducted by someone else. This allows them to identify the limitations and weaknesses of their work (sampling, procedure, statistical analysis, etc.), invite other researchers to replicate it, and verify whether their methodology is rigorous and their results reliable. They also raise new questions and hypotheses, proposing new avenues to advance knowledge related to their subject of study.

In both justice and science, from a moral and ethical standpoint, the bar must be set high at every stage of the process. We must also minimize the risk of error by determining, in each situation, which outcome is preferable or less harmful between a false negative and a false positive, and where the limits should be drawn.

Are There Situations Where a "False Positive" Is Preferable to a "False Negative"?

The answer is yes, in certain cases, a false positive is preferable to a false negative. Of course, this depends on strict, case-by-case conditions that must be based on the reliability of the situation's assessment. Let's return once again to the example of a severe viral contagion. Recall that the accuracy of a screening test is measured in two ways: (1) by how often it correctly detects people who have the disease, and (2) by how often it correctly identifies those who do not. No screening test is 100% reliable, so false positives and false negatives are inevitable. Therefore, we must decide which type of error is less harmful.

We've already pointed out that, in criminal law, it is better to make a Type II error (letting a criminal go free) than a Type I error (convicting an innocent person). But is that true in every situation?

According to Christine Nielsen, CEO of the Canadian Society for Medical Laboratory Science, "although no error is desirable [...], a false positive is preferable to a false negative in terms of public health risk, because it forces the individual to take additional precautions." These measures are obviously restrictive for the person who receives an incorrect diagnosis (isolation, treatment, etc.), but from a moral, ethical, and social standpoint, when the pathology could endanger hundreds of vulnerable lives, as is the case with epidemics, the potential benefits outweigh the inconvenience caused by these errors. That is why, in cases of doubt, individuals should undergo additional testing to confirm a diagnosis. As Nielsen explains, "Even if we have an accuracy of 99.99%, which is fantastic, it still means that some screenings need further review."

There are other domains and situations where a false positive is preferable to a false negative, for instance, in safety systems. A "false fire alarm" in a building, a "false warning" from an aircraft safety system, or a "false detection" from antivirus software on a computer containing important data are all far better than the opposite.

However, in some cases, a false positive can have disastrous consequences. One example is January 8, 2020, when Ukraine International Airlines Flight 752, a Boeing 737 NG, was mistakenly shot down by Iranian air defense shortly after takeoff from Tehran, killing all 176 passengers on board. The operator allegedly mistook it for a cruise missile. We might also recall the collapse of the Concorde overpass in Laval. It had undergone a full technical inspection in March 2005, during which no weaknesses were detected (a false negative). On September 30, 2006, the viaduct collapsed, crushing two cars and killing five occupants.

These examples show how difficult it is to assess the potential consequences of an error, whether Type I or Type II, and to determine the limits of acceptability.

*****

For readers interested in understanding in more detail how the scientific method works, please refer to Appendix 9: The Stages of the Scientific Approach.

CHAPTER 11

SAFEGUARDS IN SCIENCE

In general, people with little knowledge of science and the scientific process are unaware that there are stringent measures to minimize errors and to correct them when necessary. Although they are not 100% airtight, the number and variety of these measures make them highly effective.

Let's look at a few.

1. Ethical approval and permissions

One of the first safeguards comes before the research even begins. Once researchers have detailed their research protocol, it must be approved by an independent Ethics Committee (e.g., a Research Ethics Board), and thereafter they must obtain permission to proceed with the experiment from another official body (e.g., Health Canada's Health Products and Food Branch (HPFB)). Additional approvals may be required depending on the type of research being conducted (use of hazardous chemicals, animal products, sophisticated instruments, etc.).

2. Pre-publication review

As we have just seen, in order to publish a scientific study, a committee of experts must evaluate the researchers' work in detail (theoretical background, research objective, sampling, measurement instruments, methodology, references…). It is not uncommon for researchers to have to return to the drawing board, and sometimes their article is rejected. Authors of a paper in a specialized scientific journal are thus the first to take a critical look at their own work. In the "Discussion" section, the authors are in fact required to identify the limitations of their work themselves and to propose new avenues of research to clarify and deepen certain aspects.

Once published, the report will be read and, potentially, publicly critiqued by other researchers around the world. Publication is important for advancing knowledge and raising further questions to be explored. It is part of the long and never-ending process of acquiring knowledge.

3. Replication of studies

Everything does not end once an experiment has produced results that support or refute a hypothesis. To deepen knowledge and eventually draw more robust conclusions than those provided by a single study, science adds an essential process: replication. Replication studies are "an independent repetition of a previously published study, conducted under similar circumstances and using similar methods." That is why, in most fields, particularly in medicine, where the stakes are high, studies are replicated. Replication not only confirms the validity of the initial results, but also helps evaluate the benefit–risk ratio, as in the case of a medication, its side effects, comparisons with existing treatments or drugs… Moreover, and this is far from negligible, it helps ensure the impartiality and independence of research.

4. Meta-analysis

When a large number of studies on the same topic have been published, independent researchers often carry out what is called a meta-analysis. Stéphane Buteau summarizes it well:

"Meta-analysis is a statistical approach that allows quantitative synthesis, by calculating a combined effect, of the results of independent studies dealing with a well-defined research question. This synthesis follows a systematic review and involves a rigorous methodology intended, among other goals, to ensure the impartiality and reproducibility of the synthesis. When used appropriately, meta-analysis […] makes it possible to draw meaningful conclusions from the body of published data, thus becoming a powerful tool to support decision-making, notably in public health."

5. Market authorization and approval

When sufficient data have accumulated through the long and painstaking process of scientific research and development, an application for market authorization or approval can be submitted to the government body in each country responsible for issuing it. This process aims to certify that a product meets predetermined quality and safety standards. Its purpose is to guarantee to consumers that the product they purchase corresponds to what they are entitled to expect and that its use is safe.

6. Quality control

In virtually all areas where products that may pose risks to the public are introduced, there are very strict quality control mechanisms. This is the case, for example, in food processing, where strict standards are set, such as those of the Ministry of Agriculture, Fisheries and Food:

"To control risks related to the safety of foodstuffs, establishing and maintaining a quality system involves adopting appropriate measures. These concerns:

• specification and control of raw materials, processes, and finished products

• hygiene and sanitation policy as well as staff training

• environmental control

• recall plan

• allergen monitoring."

In pharmaceuticals, pharmacovigilance corresponds to Phase 4 of clinical trials, and it begins as soon as a drug is marketed and people begin using it as a treatment. Led by public health agencies in each country, pharmacovigilance gathers all useful information for monitoring medicines, notably rare adverse effects or those that appear only over a long period. There have been cases where drugs previously approved by a competent authority have been withdrawn from the market following significant adverse effects.

Note that this kind of vigilance is not exclusive to medicine. It also exists in many fields: technology, cosmetics, food, aviation, children's toys, etc.

7. Post-publication monitoring and vigilance

Beyond all these safeguards, one question remains: despite the rigor and controls imposed by the scientific process, is it possible that shoddy, poor-quality, or fraudulent research could slip through the protective net and be published in prestigious journals? The answer is yes, it is possible, and there are classic cases that have left a mark, including the fraudulent 1998 publication in The Lancet in which British surgeon and researcher Andrew Wakefield and his team claimed to have proved that MMR vaccines (measles, mumps, rubella) given to children could cause autism. This surprising publication alarmed the scientific community, fueled a jumble of beliefs among anti-vaccine activists, and led to policy decisions with considerable consequences. I invite you to consult Annex 1 for more details on this historic fraud, which, fortunately, was corrected.

We could cite many examples in non-scientific sectors where errors, lies, and heinous frauds took years, sometimes centuries, to be acknowledged and even longer to be corrected. The advantage of the scientific approach and its multiple safeguards lies in the fact that, in general, faults and flaws are identified and remedied more quickly.

8. Post-publication peer evaluation

As soon as scientific research is published, "post-publication" peer-review mechanisms kick in. Among the best known is PubPeer, an online platform that allows users to raise questions of scientific ethics, deontology, and integrity, leading, when necessary, to the correction or retraction of faulty articles.

In addition, all retracted articles are kept for consultation in a database, Retraction Watch. This database also allows scientists to check whether their own publications or those of their peers have been subject to post-publication evaluation.

Retraction criteria are set out by the Committee on Publication Ethics (COPE) and have highlighted shortcomings in numerous scientific studies. The authors of Retraction Watch "maintain that retractions show how science self-corrects, but can also reveal cases of scientific fraud."

A recent article published in the journal Nature notes that when Retraction Watch was created in 2010, the number of retractions averaged 45 per month. In 2021, this rose to 300 per month. The Retraction Watch database, launched in 2018, contains nearly 35,000 entries.

The mechanisms and examples presented so far show the extent of vigilance and rigor in the scientific world. In cases of doubt and mistrust, the vigilant reader can verify what they find in the bazaar of the Internet and social networks using the tools presented here.

And despite all these safeguards, even scientists can fall into the trap of so-called "predatory journals and publishers."

9. Publish or Perish and "predatory journals and publishers"

We all know competition exists in almost every field, and the scientific community is no exception! If they want to advance in their careers, researchers face a harsh reality that, for some, is truly daunting: publish or perish. To be hired by prestigious labs, promoted to tenured professorships, and obtain research grants, scientists must present a well-stocked CV showing regular publications in scientific journals. The issue is broad, and the aim here is not to present it exhaustively. Knowing that researchers and labs need to publish to survive, clever operators have found ways to exploit this vulnerability: predatory publishing.

In 2019, a group of leading academics and publishers from ten countries agreed on the following definition of this scourge:

"Predatory journals and publishers are entities that prioritize self-interest at the expense of scholarship and are characterized by false or misleading information, deviation from best editorial and publication practices, lack of transparency, and/or the use of aggressive and indiscriminate solicitation practices."

But the problem predates 2019. Between 2010 and 2017, scholarly librarian Jeffrey Beall compiled a list including 1,294 standalone journals and 1,155 publishers that had allegedly engaged in this fraudulent practice.

A Rabbit in the Clouds

In exchange for substantial sums of money, these predatory journals agree to publish works without quality control, without checking for plagiarism, and even without ethical approval.

Thus, not only can readers be duped by such publications, but researchers can be as well.

Mauro Sylos-Labini and Natalia Zinovyeva conducted a survey of 46,000 Italian scholars seeking promotion in academia. They found that about 5% had already been published in journals included on Jeffrey Beall's blacklist of "potential, possible, or probable predatory journals." Sylos-Labini and Zinovyeva refer to an analysis suggesting that predatory publishers extract millions of dollars from researchers desperate to publish.

Figure. Non-exhaustive list of safeguards in science.

*****

A Rabbit in the Clouds

The lure of profit often leads to all kinds of abuse, and industries that depend on scientific and technological research and development are, unfortunately, no exception. The number of charges and convictions involving various sectors, petrochemical, automotive, pharmaceutical, and agri-food, among others, attests to this reality.

Safeguards and oversight bodies, like laws, are not infallible.

However, as a general rule, if a fraudulent publication manages to slip past editors, post-publication reviews, literature reviews, and meta-analyses by researchers and experts, as well as replication studies,these layers of scrutiny will quickly uncover the deception and expose it so it can be corrected. And this exposure usually brings enormous consequences for the careers of the researchers involved.

CHAPTER 12

A FEW PRACTICAL TIPS FOR THE READER

Throughout history, every civilization has created and institutionalized superstitions, myths, religions, political ideologies, or pseudo-scientific and conspiratorial beliefs. Although not the only reason, such systems have played an important role in attempting to establish a "shared vision of the world" and maintain a certain "social cohesion." Conversely, the proliferation of diverse and divergent beliefs, along with the stubborn defense of traditional dogmas, has often shaken the social order and disrupted collective balance.

Once we recognize and understand these human behaviors, how can we avoid them? One suggestion is to cultivate a stronger scientific mindset. This doesn't mean we must all become laboratory researchers, but rather that we should adopt a similar approach to inquiry and reasoning. In many situations, this process depends largely on ourselves, leading to the recurring question: how to start?

Here are a few starting points and examples to guide the reader.

1. **Distinguish between coincidence, correlation, and causation.**

Our brains are remarkably good at creating connections. Many of us have inherited the mistaken belief that "there's no such thing as coincidence." Yet coincidences exist, and we experience them daily. Two events occurring close in time or space do not necessarily share a causal link.

To avoid falling into fanciful reasoning, one must verify and analyze more deeply. The first step is to determine whether a correlation exists between the factors being examined. Mathematical equations allow us to calculate the "strength" of the relationship between variables, known as the correlation coefficient. A common mistake, however, is assuming that a high correlation implies causation. It may be that both observed phenomena depend on other variables that were not included in the equation.

(See Annex 2: Factors Correlated with Life Expectancy.)

2. **Check the source of information.**

Whenever you encounter information, no matter how convincing or appealing, it is important to verify precisely where it comes from. The further an item travels from its original source, the more likely it is to be distorted. Most of us have played the "telephone game" and seen how a simple message can morph after multiple retellings. Only by tracing information back to its origin can we assess its accuracy.

3. **Validate information sources.**

"Who said what?" Referring to an authority is not inherently bad; indeed, we all do it. However, we must also ask: does the information come directly from that authority, or from a third party (journalist, documentarian, friend, influencer, YouTuber, etc.)? To answer, go back to the initial publication.

(See Annex 3: Sudden Infant Death After Vaccination for an illustrative example.)

4. **Whenever possible, consult primary sources.**

On social media or in traditional media, scientific-sounding claims abound. Usually, journalists and posters are acting in good faith, not with malicious intent. Returning to the primary source allows you to evaluate objectively whether the claim represents an author's opinion, a second-hand summary, or a properly conducted research report. Remember that all humans, including experts, have personal opinions, which should not be confused with verified facts. Even professionals can fall prey to cognitive biases, distorting the way data are presented or interpreted. Anyone who ventures beyond their area of expertise risks having to justify such overreach before their peers and the broader scientific community.

(See Annex 4: Nature Communications vs Science & Vie.)

5. **Distinguish science from pseudoscience.**

When people want to lend credibility to a belief system, they often cloak it in scientific-sounding jargon. Presentation takes precedence over substance: the packaging disguises or alters the underlying meaning. Once again, source verification becomes crucial. There are also dedicated tools, such as fact-checking organizations, that help separate authentic findings from fabricated ones.

6. **Use fact-checking resources.**

If certain claims seem exaggerated, dubious, or overly assertive, do not hesitate to investigate. Numerous fact-checking organizations examine publications, identify their sources, and assess the credibility and past record of their authors. It is best to consult several to see whether their conclusions converge.

(See Annex 5: The Truth About COVID.)

7. **Avoid abuse of language, references, and historical symbols.**

Language, the words, metaphors, and analogies we use, is the raw material from which beliefs are built. Many belief systems rely on distorted historical references to amplify emotional appeal.

The COVID-19 pandemic, for example, gave rise to a torrent of opinions, beliefs, theories, and rhetoric, alongside manipulative propaganda, sensationalism, and sometimes sheer delirium. A hallmark of conspiratorial thinking is its frequent abuse of language, historical references, and symbols.

- Abuse of language: using extreme terms such as calling public health measures a "sanitary dictatorship" or "fascist regime," or claiming society is in "collective hysteria."
- Abuse of historical references: comparing pandemic management to Nazism, lockdowns to deportation, vaccination to gas chambers or lethal injections, or accusing doctors of violating the "Nuremberg Code."
- Abuse of symbols: such as anti-vaccine protesters wearing yellow stars inscribed "unvaccinated." These analogies are not only false but profoundly indecent.

(See Annex 6: Marie-Dominique Asselin, "L'argument de l'extrême : réflexion sur la rhétorique de l'Holocauste dans les médias en temps de pandémie.")

8. Resist the temptation to rely on anecdotes.

Beliefs and theories often arise from anecdotes. According to Larousse, an anecdote is "a marginal fact, relative to one or more individuals, unpublished or little known, to which meaning may be attached but which remains incidental to the essential."

Science does not treat a single coincidence or story as proof of anything. Observations must be reproduced often enough to yield sufficient data for statistical analysis. Anecdotes blur the line between the exception and the rule.

(See Annex 7: The Exception That Creates the Rule.)

9. Be wary of irrational beliefs.

Vera Walburg, Solène Arnault, and Stacey Callahan define irrational beliefs as dysfunctional thoughts acquired in childhood through upbringing and life experience, often leading later to emotional distress.

Since the 1950s, cognitive psychology and neuroscience have shown that cognition is central to understanding and treating psychological disorders. Based on the received information, the brain builds cognitive networks that shape behavior, emotions, and how we interpret reality.

Albert Ellis, founder of Rational Emotive Behavior Therapy, identified ten categories of irrational beliefs, later grouped by other psychologists into four:

– Absolute demands: unrealistic expectations of people or events ("He never does anything right," "Things like that shouldn't happen").

– Catastrophizing: exaggerating the negative consequences of an event ("This is terrible, horrible, unbearable").

– Low frustration tolerance: inability to cope with discomfort ("I can't stand this anymore").

– Global evaluations of human worth: judging people, or oneself, as entirely worthless ("If I'm not perfect, I'm nothing").

Andreea Vîslă, Christoph Flückiger, Martin G. Holtforth, and Daniel David, from Babeș-Bolyai University in Romania, published a meta-analysis showing that irrational beliefs are associated with various forms of distress, anxiety, depression, anger, shame, and guilt.

Cognitive and cognitive-behavioral therapies (CBT) have proven highly effective in helping individuals identify, challenge, and modify such beliefs.

10. Stay alert to your own cognitive biases.

The list of cognitive biases is long and fascinating. Several have been discussed throughout this book. Overcoming them is not easy, and few solutions exist.

(See Annex 8: An Experiment on Confirmation Bias, our tendency to favor information that confirms our pre-existing beliefs and ignore that which challenges them.)

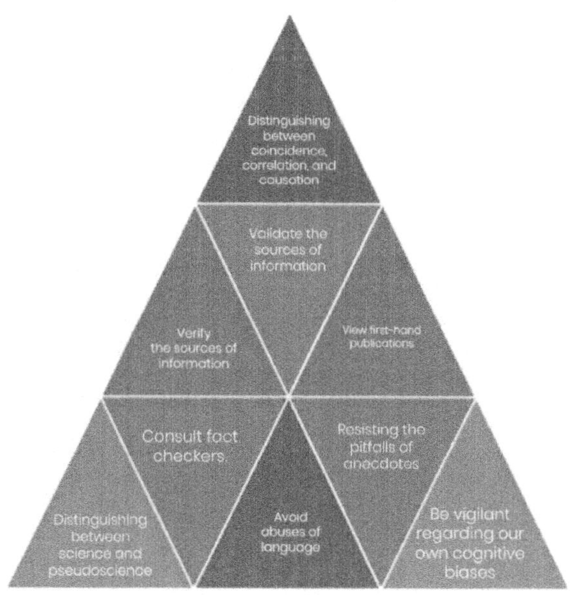

Figure. Practical advice.

A Rabbit in the Clouds

*****

It is common to emphasize the importance of developing a critical mind. This is indeed valuable, but in my view, it is not enough. The Office de la langue française defines critical thinking as "the human mind's ability to analyze and question principles taken for granted or commonly accepted ideas." Therefore, as we have seen, one must go beyond mere empirical observation and basic logical reasoning to adopt an approach that is detached from any form of subjectivity or preconceived notions. I acknowledge that this is a demanding exercise, but it is a necessary one if we wish to free ourselves from the hold that beliefs can exert over our lives.

CONCLUSION

Generally, authors reveal their motivations for writing in the opening pages of their work. In my case, I have chosen to do so only now, at the very end of this book.

You will probably not be surprised if I tell you that, as for many writers, it all began with a personal experience. The reason I chose to share it here rather than in the introduction is simple: I did not want to give my own story more importance than the underlying mechanisms from which I myself was not exempt.

Why did I write this book?

Toward the end of my adolescence, a childhood friend shared with me a few books written by a man named Claude Vorilhon, better known as Raël. In his writings, the author claimed he had been contacted by extraterrestrials in the early 1970s and that they had given him a mission: to deliver a "message to humanity" about its origins and its future. They also supposedly asked him to found a religious movement to help him fulfill that mission. Captivated by this story and everything surrounding it, I joined the Raëlian Movement and remained an active member from 1976 until 2017, over four decades. There are many reasons why I was drawn to this movement, why I stayed for so long, and why I eventually left.

After that long experience, several questions emerged in my mind, the main ones being:

- What made me adhere to that belief system?

- Why did I believe in it for so long?

- Why did I think I was different from the billions of other believers who have lived on this planet?

- Why did I feel privileged to know the truth while everyone else was supposedly mistaken?

- In short, why did I believe my convictions were more valid than those of others?

I decided to take a step back and look critically at a journey that had occupied two-thirds of my life. I wanted answers. Given my scientific training in psychology and my 35 years of experience teaching it, I could no longer settle for partial or simplistic explanations. I already possessed a wealth of knowledge from classical research in social and cognitive psychology, on topics like social learning, conformity, obedience, prejudice, cognitive functions, and biases. But I was still missing specific elements related to the construction and adherence to beliefs.

Then came the pandemic, which slowed the world down for nearly two years. Suddenly, before our very eyes, we witnessed an explosion of information, where science and pseudoscience, rigor and absurdity, reason and delirium all blended together. I wondered how a single question could produce such a vast range of contradictory answers. I also wondered whether there might be common factors behind this diversity of theories and beliefs, most of them based on no tangible evidence and resting solely on one thing: the faith of their followers. And if such factors existed, what were they?

A Rabbit in the Clouds

I began exploring the specialized literature and discovered a goldmine of scientific publications. Competent, rigorous researchers from psychology departments and research centers around the world had examined these very questions. I was struck by the depth and convergence of their findings, which helped me identify and understand many mechanisms involved in the creation of beliefs. What fascinated and reassured me most was realizing, as I read through these studies, that I could see myself, my own trajectory, my own convictions, my own engagement, reflected in them. Everything was there.

The context was ripe for such beliefs. It was the late 1970s when I first discovered Raël and his message. Like most of my generation, I grew up in a Catholic culture, catechism, Sunday Mass, prayers to the little Jesus, confessions, and so on. At the same time, Quebec society was undergoing the Quiet Revolution, a period of major social reform that separated the state from the Catholic Church. I was also fascinated by the Flower Power movement, its message of peace and love, freedom, artistic expression, and humanism, and by the rise of New Age spirituality, meditation, Eastern philosophies, the space race, humankind's first steps on the Moon, and the growing fascination with extraterrestrial life. The ingredients were all there, and Raël knew how to use them well. For me, the soil was fertile for this new religious current.

Once a few pieces of the puzzle fell into place, I began to see the world through a new lens (illusory pattern perception), to find explanations for life's mysteries (need for cognitive closure and existential motivations), and to believe that hidden powers had kept the "great truth" from us (intentionality bias). Social factors reinforced it all: being part of a select group (need for uniqueness) that possessed the truth and would "save humanity" (collective narcissism). The more I invested myself, the stronger my convictions became (confirmation bias), and the more willing I was to give everything (commitment escalation). In short, the machinery of belief formation and adherence was running at full speed.

But time does its work. Gradually, the accumulation of doubt began to erode the foundations of my faith. Slowly but surely, it weakened until one day, everything collapsed like a house of cards.

What struck me most, as I delved into the psychological research on this subject, was realizing that Raëlism exhibited all the defining characteristics of conspiratorial thinking, as summarized in the table below.

Conspiratorial Thinking	Examples from Raëlian Beliefs
Denounces an alleged conspiracy	We have been lied to about our extraterrestrial origins; we have been kept in ignorance in order to better control us…
Identifies a group of conspirators	The Catholic Church

Conspiratorial Thinking	Examples from Raëlian Beliefs
Puts forward "evidence" that appears to support its claims	The Bible, religious writings, and mysterious remains (Pyramids, Easter Island, etc.) are said to be traces left by extraterrestrials…
Falsely suggests that nothing happens by chance and that coincidences do not exist; appearances are misleading, and everything is connected	Chance does not exist; all religions have the same origin; everything is connected…
Divides the world into two camps: the good and the bad	Those who are on Yahweh's side and those on Satan's side; Raëlians as holders of the truth and others as being in error…
Designates scapegoats, which may be individuals or groups	Religious authorities, political powers, the wealthy…

It was then that I lifted my gaze beyond my own beliefs and personal experiences to look at things with clarity and honesty. I was genuinely astonished by the magnitude of the problem. I therefore set myself to work, to bring together, in a single volume, the many motives and mechanisms involved in the creation and adherence to beliefs. That is why I decided to write this book.

*****

A Rabbit in the Clouds
My Personal Message

What I hope readers will take away from this book is the essential role of the scientific mindset, especially when we are exposed to information that shapes our choices, our decisions, and our individual and collective actions. We must, of course, learn to distinguish between information that is trivial and information that truly affects our lives, the lives of those around us, our society, and the natural world.

Scientific culture begins, and largely depends on education. A great deal of excellent work is already being done in schools: playful science activities, simple laboratory experiments, hands-on demonstrations, and so on. But I also believe it is crucial to learn how to apply scientific thinking in our everyday lives, both to small and large matters. It is useful to use examples from chemistry or biology to illustrate methodology and causality, but I strongly advocate introducing more examples from the human sciences from the earliest school years, highlighting the complexity of human phenomena and clearly explaining the limits of the explanations we can offer.

The pandemic should not only have revealed the flaws in our healthcare systems, questionable political decisions, or conflicts of interest. It should also serve as a powerful source of lessons, teaching us to better:

- "Read" the information being disseminated;

- "Decode" the perceptual and cognitive biases of those who select, interpret, and distribute scientific data;
- "Distinguish" genuine scientific knowledge from pseudoscientific belief;
- "Step back" and "verify" where information comes from and what it is truly based on.

In this broad effort to better understand the human condition, we must emphasize the importance of studying how beliefs are formed; how people come to adopt them; how they influence choices, behaviors, and decisions; and what roles they have played throughout our collective history, serving various interests and enforcing systemic beliefs in countless ways.

We must uphold the fundamental right of every individual to believe what they wish, but also remain aware of the consequences that arise when personal beliefs evolve into collective doctrines prescribing behaviors, morals, and customs.

Developing a scientific culture should also help us differentiate between observation, description, correlation, and causation. History is filled with examples showing how quickly and intuitively humans interpret events and phenomena, often too quickly. It also shows how long it can take, years, centuries, even millennia, to truly explain them in depth. Between observation and explanation lies that persistent human urge to fill in the gaps, to soothe our discomfort in the face of mystery and the unknown.

A Rabbit in the Clouds

"It is precisely to remedy these weaknesses of the individual mind that science was invented (or established itself as such). To this day, it remains the only human system of thought, epistemic through its observation of reality, and social through the confrontation of theories and results, replication, peer review, and publication, that possesses the unique ability to change, improve, and correct itself, unlike any form of belief."

, Pascal Wagner-Egger

And finally, from a psychological perspective, one of the hardest lessons to learn, and to live by, is to remain serene in the face of the unknown, the ambiguous, and the mysterious. Doing so helps us ease our existential anxieties and our obsessive need to fill every void within us, the very impulse that makes us mistake a random cluster of water droplets in a cloud for the shape of a rabbit we imagine seeing there.

APPENDIX 1
POST-PUBLICATION MONITORING AND VIGILANCE

Example: The Fraudulent Publication on the Alleged Link Between Vaccines and Autism

A striking example of post-publication vigilance concerns the infamous 1998 study published in The Lancet by twelve co-authors under the supervision of British surgeon and researcher Andrew Wakefield.

This publication falsely claimed to have established a causal link between the combined trivalent MMR vaccine (measles, mumps, and rubella) and what the authors called "autistic enterocolitis," a supposed intestinal inflammatory disease allegedly responsible for the onset of autism spectrum disorders (ASD) in vaccinated children. The article was cited thousands of times and fueled the global controversy surrounding childhood vaccination.

However, numerous severe faults in Wakefield et al.'s work were later uncovered:

1. The study had serious methodological flaws; it was based on only twelve children and lacked a control group.

2. No approval from an ethics committee was ever obtained to conduct the research.

3. The autistic children involved were subjected to unjustified and invasive medical procedures (such as colonoscopies and lumbar punctures).

4. Wakefield had a clear conflict of interest: he was personally engaged in an anti-vaccination crusade and, leveraging his "medical credibility," publicly recommended that children not be vaccinated. Following this, the incidence of measles increased dramatically in the U.K. and elsewhere, as worried parents refused to immunize their children against this potentially deadly disease. Moreover, Wakefield personally profited by selling diagnostic test kits, from which he reportedly earned millions of dollars.

Meanwhile, growing doubts were raised about the study's scientific validity:

1. By 2004, no independent research had been able to replicate Wakefield's results or confirm the existence of the alleged "autistic enterocolitis."

2. That same year, ten of the twelve co-authors publicly retracted their involvement, publishing their reasons in The Lancet.

3. In 2010, Wakefield was officially sanctioned by the U.K.'s General Medical Council (GMC) and stripped of his medical license.

4. Shortly thereafter, The Lancet issued a full and permanent retraction of the 1998 study.

5. Finally, in 2014, the University of Sydney published a meta-analysis of five cohort studies involving 1.5 million children, confirming unequivocally that neither vaccines nor their components are associated with autism or autism spectrum disorders.

This case remains one of the most notorious examples of scientific fraud in modern history. It also illustrates the strength and necessity of post-publication scrutiny and peer review within the scientific community, as well as the capacity of science to self-correct, even after serious breaches of integrity.

APPENDIX 2
DISTINGUISHING COINCIDENCE, CORRELATION, AND CAUSATION

Example: Some Factors Correlated with Life Expectancy

Figure 1 clearly shows that from 1960 to 2020, life expectancy at birth increased by about twenty years worldwide (black line), rising from 52.6 years to 72.7 years.

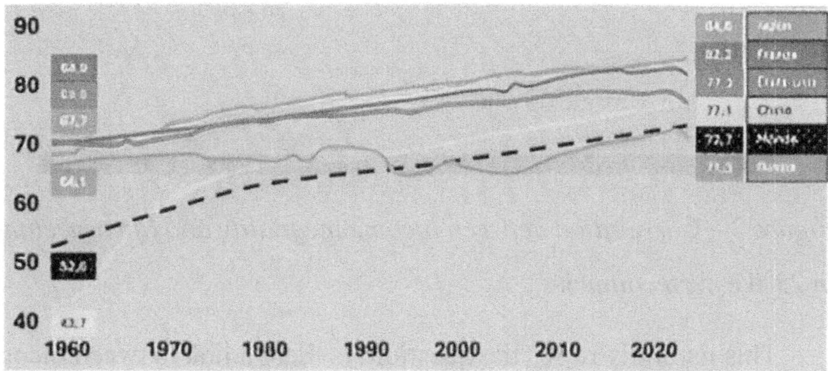

Figure 1. Evolution of life expectancy worldwide and in 5 countries, between 1960 and 2020

Figure 1 – Increase in life expectancy worldwide and in selected countries (China, United States, France, Japan, and Russia).

Several factors have contributed to this progression. For example, Figure 2 below shows a **negative correlation** between social inequality and life expectancy across 23 Western countries: in general, the greater the income inequality, the shorter the life expectancy.

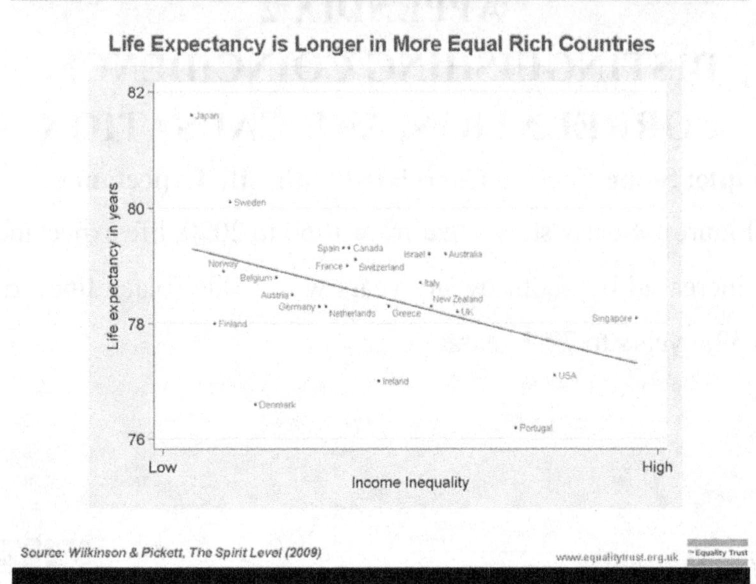

Figure 2 – Correlation between income inequality and life expectancy in 23 Western countries.

This naturally raises the question: Is there a link between income inequality and life expectancy? The answer is **yes**, the correlation shown is clear. But is it *because* of income inequality that life expectancy decreases? In other words, would increasing the average income of a population necessarily make people live longer?

The answer is **not that simple**, because income level coexists with many other factors, such as access to clean water, hygiene, and healthcare quality, early disease detection, nutrition, and more.

Looking at Figure 2, we see that Japan, the country with the lowest economic inequality, also has the highest life expectancy. Yet below the red line (the linear regression) sits Denmark, with almost as low an inequality level as Japan, but with a life expectancy nearly as low as Portugal, a country where social inequality is much higher.

Now let's consider another statistic. It is well documented that suicide affects young people more than older ones. If life expectancy tends to increase with income, should we expect lower suicide rates in wealthier countries? Figure 3, however, shows that the number of suicides per 1,000 deaths is *inversely proportional* to GDP in European Union countries.

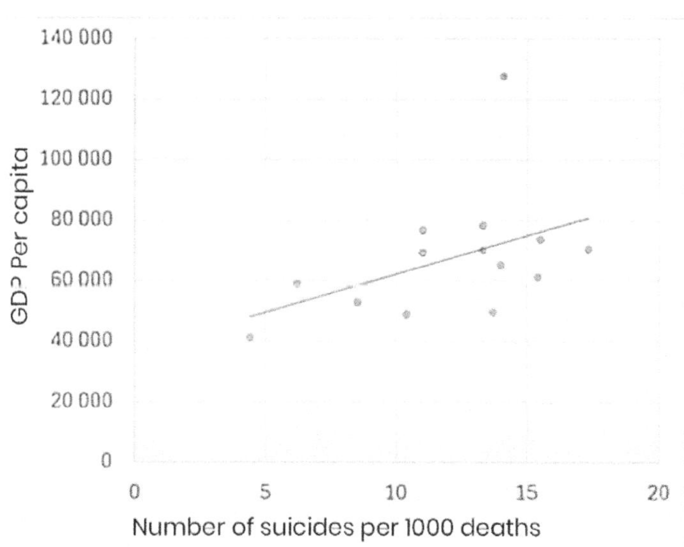

Figure 3 – Relationship between GDP and suicide rate in EU countries (number of suicides per 1,000 deaths).

Confusing correlation with causation is common. The examples above show that in the world of correlations, some relationships explain very little, and some even appear contradictory. Considering all the social advantages found in high-income countries, better education, healthcare, social services, and quality of life, one might assume that suicide rates would be lower there. Yet, as Figure 3 demonstrates, statistics are far from straightforward. Why? Because factors other than the ones being correlated often play a decisive role, and identifying them requires thorough, long-term research.

*****

Only **controlled experimental studies** can establish a true **causal link**. This is why one must be cautious of claims, often seen in the media or on social networks, that present correlations as causal relationships. In such cases, **verifying the source and reliability of the information** becomes essential.

APPENDIX 3
VALIDATING INFORMATION SOURCES

Example: Sudden Infant Death Syndrome (SIDS) Allegedly Following Vaccination

The following example concerns claims about the risks of sudden infant death syndrome (SIDS) being caused by vaccination in infants. To give some context: this controversy arose in France when vaccination against 11 diseases became mandatory for all children under two years old, born after January 1, 2018. This decision provoked a public outcry among opponents of mandatory vaccination.

As early as 2017, Dr. Luc Montagnier, Nobel Prize winner in Medicine (2008), participated in a conference titled "Yes to Vaccines, No to 11 Vaccines for Infants." During this event, Dr. Montagnier and his colleague Dr. Henri Joyeux declared:

"We are here to sound the alarm, to alert the whole country and the world. I want to raise awareness about sudden infant death. It's something terrible, the cause is unknown, but there are scientific facts showing that many of these deaths occur after vaccination..."

In early January 2018, Montagnier and Joyeux doubled down with a new open letter titled "11 Vaccines: An Urgent Call to Health Professionals," again warning of alleged dangers of vaccination in infants.

These statements and alarms were quickly picked up and circulated by many anti-vaccination groups and websites, including unpeudairfrais.org, which supported its claims with a link to a series of graphs from a paper published in the International Journal of Environmental Research and Public Health.

Let's examine several aspects related to this example:

- Montagnier and Joyeux stated: "There are scientific facts showing that many of these deaths occur after vaccination." But what are these "scientific facts"?

→ No publication was cited to support this claim.

What did exist, however, was a massive British study published in 2001, covering 17.7 million infants, which thoroughly examined the link between vaccination and sudden infant death.

The results showed no temporal association between vaccination and SIDS, and that vaccination does not cause sudden or unexpected death in infancy. On the contrary, the data suggested a protective correlation.

- Have the authors of that British Medical Journal article ever had to issue corrections or retractions?

→ A review of PubPeer and the Retraction Watch database shows no such record.

- A review of the scientific literature also confirms that no research has ever demonstrated that vaccination increases the risk of sudden infant death.
- What about the source cited by unpeudairfrais.org?

A Rabbit in the Clouds

→ The referenced paper, "Relative Incidence of Office Visits and Cumulative Rates of Billed Diagnoses Along the Axis of Vaccination," had to issue corrections on January 22, 2020.

→ Later that same year, on July 22, the journal decided to fully retract the article.

→ The authors' errors were noted publicly on PubPeer and listed in Retraction Watch.

This case is one among many in which fraudulent or misleading sources have been used to defend unfounded beliefs, with potentially disastrous consequences.

Everyone can make mistakes or unintentionally share poor-quality sources, and sometimes personal opinions are passed off as facts. But when dealing with topics as important as public health, nothing should be taken lightly. It is every author's responsibility to ensure the accuracy, validity, and integrity of the sources they cite.

APPENDIX 4
CONSULTING PRIMARY SOURCES

Example: Nature Communications vs Science & Vie

In Chapter 1, I presented the results of a neuroscience study by Alexandre Salvador et al. The article, published in Nature Communications, was titled "Premature commitment to uncertain decisions during human NMDA receptor hypofunction."

A few months later, the popular science magazine Science & Vie reported on this research. The article, written by Lucile André, was titled: "False beliefs may be caused by the dysfunction of a synaptic receptor."

This title from Science & Vie is a perfect example of a syllogistic error, similar to those discussed in Chapter 7:

"If it rains, the football game is canceled.

Today's game is canceled; therefore, it must be raining."

Such reasoning mistakes are extremely common across many fields.

The original Nature Communications study found that participants who were administered ketamine, a substance known to temporarily reduce the function of NMDA synaptic receptors, tended to make premature decisions based on ambiguous visual information, compared to those who received a placebo.

A Rabbit in the Clouds

However, Lucile André and Science & Vie jumped to an unsupported conclusion by asserting that "false beliefs are caused by synaptic dysfunction."

By consulting the primary source, the original article published by the researchers, we understand that, at this stage of knowledge, the results do not allow us to claim that false beliefs (such as conspiracy, religious, or pseudoscientific beliefs) are caused by synaptic dysfunction.

It will take years of further research to confirm or refute any causal link between NMDA receptor functioning and the formation of beliefs. For now, these findings represent interesting hypotheses, not established facts.

APPENDIX 5
DO NOT HESITATE TO CONSULT FACT-CHECKERS

Example: The Truth About COVID

On April 22, 2022, Russell L. Blaylock published an editorial titled "COVID UPDATE: What is the Truth?" in the journal Surgical Neurology International.

Aside from the somewhat presumptuous title, the very first sentence of the article is already alarming:

"The COVID-19 pandemic is one of the most manipulated infectious disease events in history, characterized by official lies in an endless stream, led by government bureaucracies, medical associations, medical boards, the media, and international agencies."

This sweeping claim is followed by three references, which I decided to check.

- The first reference is a book co-authored by Blaylock himself, titled The Chinese Virus: What Is the Truth?
- The second reference is simply an earlier version of his own editorial.
- The third comes from an anti-vaccine blog.

Despite these red flags, I continued reading. Yet, the article provided no legitimate scientific references to support any of its assertions. I therefore turned to fact-checking platforms for verification.

A Rabbit in the Clouds

PubPeer featured a reviewer's comment:

"This bizarre editorial has gone viral, so I thought it would be useful to examine some of its claims and how remarkably unscientific the entire piece is."

Multiple flaws were noted. Another of Blaylock's articles was also flagged and discussed critically on PubPeer.

MedPage Today reached similar conclusions:

- "Although the editorial was published in a peer-reviewed scientific journal, it provides no solid scientific evidence to support its claims."
- "The most frequently cited reference in the editorial is anti-vaccine advocate Robert F. Kennedy Jr.'s book The Real Anthony Fauci: Bill Gates, Big Pharma, and the Global War on Democracy and Public Health, a source that is 'obviously unreliable.'"
- The report also pointed out financial conflicts of interest, noting that "Blaylock is currently selling dietary supplements called Brain Repair Formula."

Genetic Literacy Project also criticized Blaylock for spreading misinformation and promoting his own treatments for profit.

The outlet The Outline was even harsher, labeling Blaylock as "a charlatan exploiting public fear," citing several of his previous conspiratorial claims:

- Alleging that elderly patients were euthanized under Obamacare;
- Claiming on InfoWars that fluoride in drinking water is an eugenics program causing ADHD;
- Suggesting that poor nutrition is part of an Illuminati plan to make people violent and reduce the global population;
- Reviving the fraudulent myth of a causal link between vaccines and autism in children;
- And describing vaccination as a "medical genocide" against patients.

There are many other fact-checkers that have evaluated this topic and others like it.

The takeaway is simple: always verify bold or sensational claims through multiple reliable fact-checking sources before forming an opinion.

APPENDIX 6
ABUSE OF LANGUAGE

Example: "The Argument of the Extreme: Reflection on Holocaust Rhetoric in the Media During the Pandemic", Text by Marie-Dominique Asselin

The Wearing of the Star of David and the Jews During the Holocaust

"Over the past several months, many comparisons have been made to denounce the measures implemented by the Quebec government to fight the COVID-19 pandemic. Premier François Legault has, unsurprisingly, been compared to Adolf Hitler.

The most common analogy circulating on social media during the pandemic has been the comparison between the government's mask mandate and the obligation imposed on Jews under the Nazi regime to wear the yellow Star of David.

From 1939 to 1945, Nazi Germany made it its mission to eradicate every trace of European Jewry. For this planned extermination of millions of Jews to be possible, the Nazis first needed to create a sense of *otherness* among the German population and those in occupied territories toward the Jews. One of the first measures implemented was the marking of Jews.

Before segregating them into ghettos and deporting them to camps, the Nazis required all Jews aged 12 and older to wear the Star of David. The purpose was simple: to identify the Jew, the *other*, the *different*, the *enemy*.

At first glance, there may seem to be a superficial resemblance between the two: both the face mask and the Star of David alter one's physical appearance and are mandated by authorities. In both cases, failure to comply carries consequences: death in the case of Jews who failed to wear the star, or, at worst, fines or being denied entry to certain places, in the case of mask mandates.

Beyond these vastly unequal penalties, the comparison between the Nazi-imposed Star of David and the public health requirement to wear a mask is not only illogical, but it is *grossly inappropriate and indecent*.

During the Second World War, the marking of Jews served to identify the enemy and distinguish them from authorities and other members of society. Beyond the Star of David, the Nazis used multiple markings to classify 'undesirable' individuals. At Auschwitz, for example, Jewish prisoners wore the Star of David; homosexuals were forced to wear a pink triangle; political prisoners, a red triangle; and criminals, a green triangle.

The goal of these symbols was to compartmentalize individuals into subgroups, to reduce them to the smallest common denominator. This marking system allowed Nazi leaders to identify prisoners but also to divide them, maintaining greater control over them.

A Rabbit in the Clouds

Today, in Quebec, mask-wearing is mandatory for everyone, civilians and state workers alike. Whether one agrees or disagrees with its enforcement or effectiveness, mask-wearing is not a means of dividing society or labeling specific individuals.

On the contrary, if the purpose of mask-wearing is primarily sanitary, its effect is rather to *unify* society, including state employees, rather than to stigmatize any particular group."

, Marie-Dominique Asselin, PhD in History, 2020.

APPENDIX 7
RESISTING THE TRAP OF RELYING ON ANECDOTES

Example: The Exception That Proves the Rule

In 2020, a friend shared on his Facebook page a link that was widely circulating in conspiracy circles:

"Swabs = Danger! Protect your blood-brain barrier,"

Where the author made the following alarming claim:

"Your brain, with its many vital functions, is protected from potentially harmful substances in the bloodstream by a fence-like structure called the blood-brain barrier. Why bring this up now, more than ever? Simply because the EXACT spot where the 'testers' (using the long Q-tip for the RT-PCR test sample) obtain your sample is called the blood-brain barrier."

Many comments followed this post. I decided to add my own, pointing out the numerous flaws already identified in that misleading warning.

In 2021, the same friend sent me another link, this time, to a statement issued by the French National Academy of Medicine (ANM), titled:

"Nasopharyngeal swabs are not without risk."

He added, somewhat triumphantly:

A Rabbit in the Clouds

"A few months ago, you dismissed this information I shared on Facebook. Don't you see now that you were wrong? This article proves it. I'm sending it to you so you can reconsider."

The ANM statement did note that "serious complications have begun to appear in the medical literature in recent weeks, notably fractures of the anterior skull base associated with meningitis risk." It also reminded practitioners to take proper precautions when performing PCR tests.

Once again, this information was distorted and misused by opponents of public health measures, particularly anti-testing and anti-vaccine activists, who claimed that swabs could "reach the brain" and cause hidden neurological damage, using this ANM warning as supposed proof. Clearly, those who propagated these claims did not read the ANM statement carefully, nor did they check the three case studies it referenced.

Let's examine these references.

These were all case reports, meaning detailed descriptions of individual patients' diagnoses, treatments, and outcomes. Such reports are useful for documenting rare or unusual events, but they cannot establish causality.

In one cited case, Sullivan et al. wrote:

"This patient had a previously undiagnosed skull base defect at the ethmoid fovea, visible on imaging dating back to 2017. We therefore theorize that the swab itself did not breach the skull base, but that the invasive test made contact with a pre-existing encephalocele."

They concluded:

"This case of iatrogenic CSF leak from a nasal swab for COVID-19 illustrates that prior surgery or pathology affecting nasal anatomy may increase the risk of adverse events associated with nasal testing for respiratory pathogens, including COVID-19."

Another cited study, by Föh B. et al., analyzed 3,083 patients who collectively underwent 11,476 nasal and oropharyngeal swabs. They reported:

"We observed a total of three adverse events (0.026%). In two cases, the swab tip broke off but was retrieved without complication; in one case, a patient developed spontaneous dislocation of the left temporomandibular joint while opening the mouth for the oropharyngeal swab. No nosebleeds requiring medical attention or serious adverse events occurred."

Their conclusion:

"Our results from a large, closely monitored cohort show that combined nasal and oropharyngeal swabbing is generally safe. Adverse events are extremely rare, and serious complications are unlikely, though they cannot be completely ruled out."

As Dr. Jean-Marc Juvanon, ENT specialist and member of the French Society of Otorhinolaryngology, noted:

A Rabbit in the Clouds

"Out of billions of tests performed over two years, only a very, very small number of serious cases have been reported."

This example clearly demonstrates that anecdotes cannot serve as proof for constructing a reliable theory of cause and effect. At best, case studies,when conducted rigorously,can raise a red flag and prompt temporary caution (such as suspending a drug or adjusting procedures) while more comprehensive research is carried out.

APPENDIX 8
BEING VIGILANT ABOUT OUR OWN COGNITIVE BIASES

Example: The Confirmation Bias

Two researchers from Pennsylvania State University conducted an experiment to test confirmation bias as a mechanism underlying the persistence of misinformation about climate change, and to examine the concept of attitude certainty, that is, the subjective feeling of confidence or conviction a person has regarding their own attitudes.

Their findings revealed a strong confirmation bias: individuals tend to accept and reinforce messages that align with their pre-existing beliefs, further strengthening those convictions. The researchers also found that attitude certainty acts as a driving force in belief polarization, deepening divisions between opposing viewpoints.

The primary takeaway from this study is a practical one: we must train ourselves to recognize our own cognitive biases, especially when we feel certain about something for which we lack concrete evidence.

APPENDIX 9
THE STEPS OF THE SCIENTIFIC METHOD

To illustrate the steps of the scientific method, let's take the example of pharmaceutical research and development. This topic, which has generated much ink and debate, lends itself perfectly to this exercise.

The development and marketing of a new drug or vaccine were widely discussed during the COVID-19 pandemic, and we have read, seen, and heard almost everything on the subject. I suggest setting aside the debates and controversies surrounding the pharmaceutical industry, though they are interesting and relevant from financial, legal, and even moral standpoints, and focusing here strictly on the scientific aspect. The following example deals with pharmacological research and development, but it is not exclusive to this field. In the scientific world, researchers are bound to apply very strict rules throughout this long and sometimes perilous process, which often spans many years.

Scientific Steps in the Research and Development Process

The following figure illustrates the steps of the scientific method through medical research.

A Rabbit in the Clouds

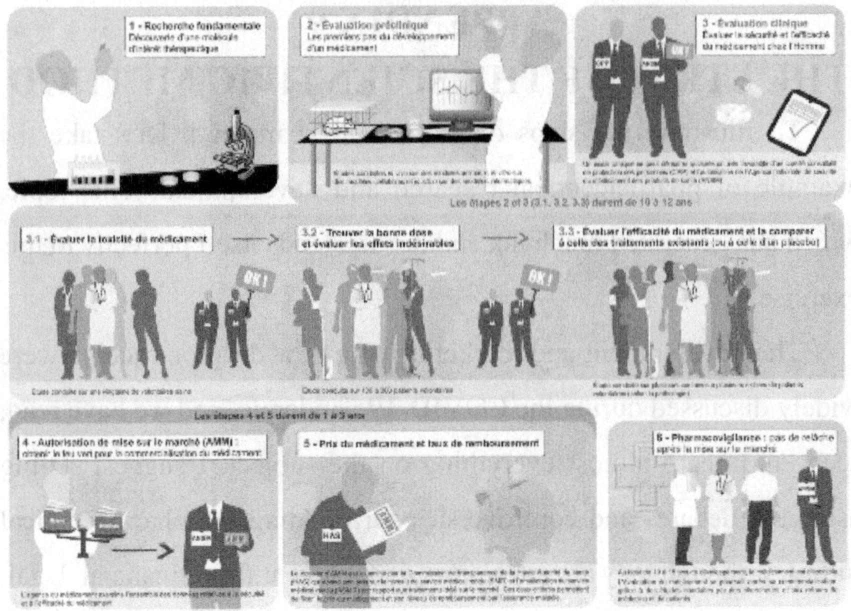

© INSERM

Figure. Illustration of the stages in the research and development of a new drug

For both small and large-scale research projects, the scientific process always begins with a question. Whether seeking solutions or answers to problems in mechanics, health, psychology, chemistry, or astronomy, questions are asked, and these questions trigger a process intended to answer them.

In our example, the first step is to find molecules that may have therapeutic potential. It's important to note that, on average, "out of 10,000 screened molecules, 10 may be patented, and only one (1) will pass all testing and clinical trial stages to become a drug, the path from innovation to patient is long (around twelve years on average), complex, and costly."

A Rabbit in the Clouds

The second step, called the preclinical phase, aims to identify a therapeutic target involved in a disease and to design various models (in vitro, in vivo, and in silico) of the disease and its treatment.

Next come the three phases of the third step, clinical trials:

- Phase 1 assesses the toxicity, tolerance, and pharmacokinetics of each selected molecule. This phase is conducted on a small sample of healthy volunteers.

- Phase 2 evaluates the drug's effectiveness for the targeted disease, determines the optimal dosage, and identifies side effects.

- Phase 3 is the most demanding. Its goal is to study the drug's efficacy and safety, an indispensable step for eventual market approval.

Let's take a closer look at the methodological aspect of this third phase.

The Goal of the Research

The objective of research is to determine whether a given molecule safely contributes to treating a particular disease. In other words, researchers want to know whether a cause-and-effect relationship exists between taking the drug (the cause) and improving the disease (the effect, measurable, for example, by a reduction or disappearance of symptoms). To verify this causal link, researchers must conduct an experiment following very clearly defined steps.

These steps are the same regardless of the research domain.

The Steps of Experimentation

1. Sampling and Group Assignment

A Rabbit in the Clouds

First, researchers must recruit a sample of participants. In our case, these are individuals with the targeted disease. Recruitment must consider many pre-established factors such as age, health status, comorbidities, and any treatments that might interact with the experimental one.

Once recruited, participants are randomly and evenly divided into three groups:

The experimental group, which receives the treatment.

The placebo group receives a "fake" drug resembling the real one but containing no active substance.

A control group, which receives no treatment at all.

This step is crucial to avoid biased samples or uneven group distributions.

2. Experimental Design and Controls

Researchers must also determine which experimental design best answers their question. The choice depends on several criteria, such as the number and nature of variables under study and the research problem itself.

In this example, it is a unifactorial experimental design with independent groups, experimental, placebo, and control. The goal is to determine whether the tested drug (the independent variable) has beneficial effects on the disease (the dependent variable).

Researchers must first isolate and introduce the independent variable ("the drug") while controlling a range of controlled variables that might interfere with treatment or skew results. These include spontaneous recovery, other diseases, additional treatments, allergies, fatigue, stress, diet, and more.

Hence, during this phase, all participants are monitored under controlled laboratory conditions, including meals, sleep schedules, drug intake, and blood tests, to limit external influences.

A particularly important factor to control is the placebo effect. Research has shown that about 30% of patients given a placebo (believing it to be an effective drug) show positive psychophysiological results. Therefore, participants must not know whether they belong to the experimental or placebo group, and those administering the treatments must not know either.

This is known as a "randomized double-blind study":

Randomized because participants are randomly assigned to groups.

Double-blind because neither the participants nor experimenters know who receives what.

This procedure prevents bias and ensures reliability. The entire process is overseen by the principal investigator.

3. Data Analysis and Interpretation of Results

One can easily imagine the vast amount of data collected from hundreds or thousands of subjects, often across multiple research centers worldwide. For example, if ten measurements are taken before, during, and after the study for 1,000 subjects, that yields over 30,000 data points.

These are first compiled to perform descriptive statistics, which provide averages and distributions. However, these averages alone cannot determine whether a treatment works. Researchers must then use hypothesis testing to assess whether observed differences are statistically significant.

Scientific rigor thus operates on two levels:

- Methodological rigor, how the experiment is designed and executed;
- Statistical rigor, how data are analyzed to determine the probability of a cause-and-effect link.

The Hypothesis Test

The process resembles a criminal investigation. In criminal law, a suspect is presumed innocent, and prosecutors must provide proof "beyond a reasonable doubt" to establish guilt. This prevents wrongful convictions, a false positive, or Type I error, which must be avoided at all costs.

Similarly, in science, the goal is not to prove the hypothesis true, but to test whether evidence supports rejecting the null hypothesis.

Researchers conduct a hypothesis test, choosing between:

- The null hypothesis (H_0) assumes the drug has no effect (like the presumption of innocence);
- The alternative hypothesis (H_1) assumes the drug does have an effect.

To accept H_1, one must reject H_0. In science, it's preferable to make a Type II error (wrongly rejecting H_1, missing a true effect) than a Type I error (wrongly accepting a false positive). In practice, it's safer to wrongly conclude that a drug is ineffective than to wrongly claim it works.

A Rabbit in the Clouds

Since no research can ever be 100% error-free, science defines an acceptable margin of error, usually below 5%. This means the probability that results are due to chance (rather than the studied variable) must be under 0.05. If so, the results are considered statistically significant.

Example: Salvador et al., Nature Communications

Let's revisit the study by Salvador et al. in Nature Communications (see Chapter 1). The researchers examined whether blocking NMDA synaptic receptors with ketamine affected participants' decision certainty during ambiguous-choice tasks.

They compared several parameters, including decision uncertainty (dependent variable) under ketamine and placebo (independent variable).

The authors reported:

"Participants reversed their decisions much more frequently under ketamine than under placebo (ketamine: $42.9 \pm 2.5\%$; placebo: $34.4 \pm 2.0\%$; $p < 0.001$). This finding suggests a higher level of decision uncertainty under ketamine."

The test indicated less than one chance in a thousand that this difference was due to other factors (e.g., chance, methodological error, or individual differences). Thus, the null hypothesis (H_0) was rejected, and the alternative hypothesis (H_1) was accepted.

4. Communication of Results

Finally, researchers must publish their results in a specialized journal. Here again, there are strict criteria regarding both form and substance.

A Rabbit in the Clouds

Before publication, every scientific report must undergo peer review, an evaluation by experts in the same field, to ensure methodological, ethical, and analytical rigor before the findings are accepted and shared with the scientific community.

BIBLOGRAPHY

AARON C. Kay, Jennifer A. Whitson, Danielle Gaucher, Adam D. Galinsky (2009). Compensatory Control: Achieving Order Through the Mind, Our Institutions, and the Heavens, *Current Directions in Psychological Science,* Vol. 18, Issue 5, pp. 264-268. (https://doi.org/10.1111/j.1467-8721.2009.01649.x).

ABERTH, John (2005). *Medical Responses. In: The Black Death. The Bedford Series in History and Culture.* Palgrave Macmillan, New York. (https://doi.org/10.1007/978-1-137-10349-9_4) (Consulté le 28 octobre 2022).

ALLAIN, P. (2013). La prise de décision : aspects théoriques, neuroanatomie et évaluation. *Revue de neuropsychologie*, 5, 69-81. (https://doi.org/10.1684/nrp.2013.0257). (Consulté le 18 avril 2024).

AMERICAN PSYCHIATRIC ASSOCIATION, *DSM-5: diagnostic and statistical manual of mental disorders*, 5ᵉ édition, Washington D.C. American Psychiatric Association, 2013 (ISBN 9780890425541).

ASSELIN, Marie-Dominique (2020). *L'argument de l'extrême : réflexion sur la rhétorique de l'Holocauste dans les médias en temps de pandémie.* Dans : Histoireengagée.ca, 20 décembre 2020. https://histoireengagee.ca/largument-de-lextreme-reflexion-sur-la-rhetorique-de-lholocauste-dans-les-medias-en-temps-de-pandemie/ (Consulté le 21 novembre 2022).

ASTOLFI Jean-Pierre, DAROT Éliane, GINSBURGER-VOGEL Yvette *et al.*, « Chapitre 3. Conflit cognitif, conflit socio-cognitif », dans : *Mots-clés de la didactique des sciences. Repère, définitions, bibliographies*, sous la direction de ASTOLFI Jean-Pierre, DAROT Éliane, GINSBURGER-VOGEL Yvette *et al.* Louvain-la-Neuve, De Boeck Supérieur, « Pratiques pédagogiques », 2008, p. 35-48. (https://www.cairn.info/mots-cles-de-la-didactique-des-sciences--9782804157166-page-35.htm) (Consulté le 29 avril 2024).

AUSMAN J. I., Blaylock R. L. *The China Virus. What is the truth?* United States: James I. and Carolyn R. Ausman Education Foundation (AEF); 2021.

AVEZUM, Álvaro & al. Hydroxychloroquine versus placebo in the treatment of non-hospitalised patients with COVID-19 (COPE – Coalition V): A double-blind, multicentre, randomised, controlled trial, The *Lancet Regional Health - Americas, Vol. 11*, 2022, 10024. (https://doi.org/10.1016/j.lana.2022.100243).

BAR-HEN, Avner (2020). Statistiques : l'apophénie, ou comment donner du sens à ce qui n'en a pas. Dans : https://www.sciencesetavenir.fr/fondamental/mathematiques/statistiques-l-apophenie-ou-comment-donner-du-sens-a-ce-qui-n-en-a-pas_37881. (Consulté le 29 mai 2022).

BARRY, John M. (2017). How the Horrific 1918 Flu Spread Across America, | History | *Smithsonian Magazine, nov. 2017.* (https://www.smithsonianmag.com/history/journal-plague-year-180965222/)

BÉNA, J., Carreras, O. & Terrier, P. (2019). L'effet de vérité induit par la répétition : Revue critique de l'hypothèse de la familiarité. *L'Année psychologique, 119*, 397-425.

BERGER, J., & Milkman, K. L. (2012). What Makes Online Content Viral? *Journal of Marketing Research, 49(2)*, pp. 192–205. (https://doi.org/10.1509/jmr.10.0353).

BERNS, G. S., Bell, E., Capra, C. M., Prietula, M. J., Moore, S., Anderson, B., Ginges, J., & Atran, S. (2012). The price of your soul: neural evidence for the non-utilitarian representation of sacred values. *Philosophical transactions of the Royal Society of London. Series B, Biological sciences, 367*(1589), 754–762. (https://doi.org/10.1098/rstb.2011.0262)

BLAYLOCK R. L. (2021). Covid-19 pandemic: What is the truth? *Surgical neurology international, 12,* 591. (https://doi.org/10.25259/SNI_1008_2021)

BLAYLOCK R. L. (2022) COVID UPDATE : What is the truth? *Surg Neurol Int*, 13 (167). (https://surgicalneurologyint.com/surgicalint-articles/covid-update-what-is-the-truth) (Consulté le 7 novembre 2022).

BOUCHER, O., Citherlet, D., Ghaziri, J., Hébert-Seropian, B., Von Siebenthal, Z. & Nguyen, D. (2017). Insula : neuropsychologie du cinquième lobe du cerveau. *Revue de neuropsychologie*, 9, 154-161. (https://doi.org/10.1684/nrp.2017.0422).

BOURQUIN, Jimmy (2020). 1918-1920 - Quand un médecin prétendait avoir trouvé un remède contre la grippe espagnole, *RadioFrance*, 2 juillet 2020. https://www.radiofrance.fr/franceinter/1918-1920-quand-un-medecin-pretendait-avoir-trouve-un-remede-contre-la-grippe-espagnole-5155507 (Consulté le 28 décembre 2022).

BOUVARD, Martine (2013). Les cinq dimensions de la personnalité. *L'Essentiel Cerveau et Psycho*, N°16 - Novembre 2013. (https://www.cerveauetpsycho.fr/sd/psychologie-comportementale/les-cinq-dimensions-de-la-personnalite-7652.php) (Consulté le 18 juin 2022).

BRADLEY, Franks, Adrian Bangerter et Martin W. Bauer (2013). Conspiracy theories as quasi-religious mentality : an integrated account from cognitive science, social representations theory, and frame theory, *Frontiers in Psychology, 16* (4), p. 424. (doi : 10.3389/fpsyg.2013.00424. PMID : 23882235; PMCID : PMC3712257).

BRADY, W. J., Gantman, A. P., & Van Bavel, J. J. (2020). Attentional capture helps explain why moral and emotional content go viral. *Journal of Experimental Psychology : General, 149*(4), 746–756. (https://doi.org/10.1037/xge0000673).

BRASHIER, N. M. and E. J. Marsh (2020). *Judging Truth*, Annual Review of Psychology, Vol. 71, pp 499-515. (https://doi.org/10.1146/annurev-psych-010419-050807).

BUTEAU, Stéphane (2016). *La méta-analyse : bien plus que le simple calcul d'un effet combiné !* Dans https://www.inspq.qc.ca/bise/la-meta-analyse-bien-plus-que-le-simple-calcul-d-un-effet-combine (Consulté le 16 septembre 2022).

CARIGNAN, Marie-Ève, David Morin, Marie-Laure Daxhelet, Sylvain Bédard, Olivier Champagne-Poirier, Emmanuel Choquette, Guilhem Aliaga, Yanni Khennache, Esaie Kuitche Kamela, LE MOUVEMENT CONSPIRATIONNISTE AU QUÉBEC LEADERS, DISCOURS ET ADHÉSION, Comprendre pour mieux agir. « Chaire UNESCO en prévention de la radicalisation et de l'extrémisme violents » Université de Sherbrooke, Université Concordia et Université du Québec à Montréal, Juin 2022.

CARREIRO, S., Smelson, D., Ranney, M., Horvath, K. J., Picard, R. W., Boudreaux, E. D., Hayes, R., & Boyer, E. W. (2015). Real-time mobile detection of drug use with wearable biosensors: a pilot study. *Journal of medical toxicology: official journal of the American College of Medical Toxicology*, *11*(1), 73–79. (doi.org/10.1007/s13181-014-0439-7)

CHATELAIN, Yannick, Caroline Cuny et Jamel Khenfer (2021). Pandémie et restrictions : quelles stratégies pour reprendre le contrôle de sa vie ? *The Conversation,* mars 2021. (https://theconversation.com/pandemie-et-restrictions-quelles-strategies-pour-reprendre-le-controle-de-sa-vie-157827) (Consulté le 10 juin 2022).

CHRISTENSEN, Larry B. (1985). *Experimental Methodology,* 3th édition, Allyn & Bacon, 544 pages.

CICHOCKA A, Marchlewska M, Golec de Zavala A, Olechowski (2016). 'They will not control us' : Ingroup positivity and belief in intergroup conspiracies, *British Journal of Psychology,* 107(3), pp. 556-76. (https://doi.org/10.1111/bjop.12158).

CICHOCKA, Aleksandra, Marta Marchlewska, Agnieszka Golec de Zavala (2016). Does Self-Love or Self-Hate Predict Conspiracy Beliefs? Narcissism, Self-Esteem, and the Endorsement of Conspiracy Theories, *Social Psychological and Personality Science,* Vol. 7 (2), pp. 157-166. (https://doi.org/10.1177%2F1948550615616170).

COMMUNIQUÉ par INSERM (Salle de Presse), 17 JANV. 2022. (https://presse.inserm.fr/un-recepteur-synaptique-implique-dans-lemergence-de-croyances-aberrantes/44508/?fbclid=IwAR3qcu2gPpsTJn82wpK6O00hN3d IKpsL39y49gUgU6lIYWv4U8lPbJMevwY) (Consulté le 6 janvier 2023).

LEVINE, T. R. (2022). Truth-default theory and the psychology of lying and deception detection. *Current opinion in psychology, 47.* (https://doi.org/10.1016/j.copsyc.2022.101380).

DARSHANI, Jai Kumareswaran (2014). The Psychopathological Foundations of Conspiracy Theorists, A thesis submitted to Victoria University of Wellington in fulfilment of the requirements for the degree of Doctor of Philosophy in Psychology. (https://researcharchive.vuw.ac.nz/xmlui/bitstream/handle/10063/3603/thesis.pdf?sequence=2).

DE ZAVALA, A. G. Cichocka, A., Eidelson, R., & Jayawickreme, N. (2009). Collective narcissism and its social consequences. *Journal of Personality and Social Psychology,* 97(6), pp. 1074–1096. (https://doi.org/10.1037/a0016904).

DE ZAVALA, Agnieszka Golec, Dorottya Lantos (2020). Collective Narcissism and Its Social Consequences : The Bad and the Ugly, Sage Journal, Vol 29, Issue 3. (https://doi.org/10.1177/0963721420917703).

DIGIUSEPPE, Raymond A., and others, 'The Irrational and Rational Beliefs', *A Practitioner's Guide to Rational-Emotive Behavior Therapy,* 3 edn (New York, 2013; online edn, Oxford Academic, 1 Jan. 2015), (https://doi.org/10.1093) (Consulté le 16 avril 2024).

DUCHESNE, Marie-Claude Montréal au temps de la variole, La Sentinelle, dans : https://www.lasentinelle.ca/montreal-au-temps-de-la-variole/. (Consulté le 7 juillet 2022).

ELUÈRE, M. & Héas, S. (2017). Superstitions, cultures et sports, entre croyances et rationalisations. Le cas exploratoire d'une équipe féminine professionnelle de volleyball en France. *Les Cahiers Internationaux de Psychologie Sociale,* 113, 25-55. (https://doi.org/10.3917/cips.113.0025).

ERB Hans-Peter et Susanne Gebert, La Lady Gaga en nous, dans *Cerveau & Psycho, N° 65*, sept. 2014. https://www.cerveauetpsycho.fr/sd/psychologie-sociale/la-lady-gaga-en-nous-8098.php (Consulté le 5 octobre 2022).

FANSI, A., Jehanno, C., Lapalme, M., Drapeau, M., & Bouchard, S. (2015). Efficacité de la psychothérapie comparativement à la pharmacothérapie dans le traitement des troubles anxieux et dépressifs chez l'adulte : une revue de la littérature. *Santé mentale au Québec*, *40*(4), 141–173. (doi.org/10.7202/1036098ar).

FAROOQ Kazi, Ammara Mushtaq (2022), Luc Montagnier, *The Lancet,* VOL. 22(4), P. 458.

FAUCHER, Olivier (2021). L'étoile jaune « est là pour rester » dans les manifestations, *Journal de Montréal, 17 août 2021.* https://www.journaldemontreal.com/2021/08/17/manifestation-contre-le-passeport-vaccinal-a-laval (Consulté le 21 novembre 2022).

FENSCH-CARRAT, Léa. La désinformation sur internet à propos des vaccins et le rôle du pharmacien d'officine dans la réinformation. *Sciences pharmaceutiques*. 2019. dumas-02355511.

FERGUSON, M. A., Nielsen, J. A., King, J. B., Dai, L., Giangrasso, D. M., Holman, R., … Anderson, J. S. (2018). Reward, salience, and attentional networks are activated by religious experience in devout Mormons. *Social Neuroscience, 13*(1), 104–116. (https://doi.org/10.1080/17470919.2016.1257437).

FISKE, S. & Taylor, S. « Comportement et cognition », dans : *Cognition sociale. Des neurones à la culture*, sous la direction de FISKE Susan T, TAYLOR Shelley E. Wavre, Mardaga, « PSY-Individus, groupes, culture », 2011, p. 417-456. URL : https://www.cairn.info/--9782804700355-page-417.htm. (Consulté le 7 août 2022).

FLEMING, P J, Blair P S, Platt M W, Tripp J, Smith I J, Golding J et al. The UK accelerated immunisation programme and sudden unexpected death in infancy: case-control study, *BMJ* 2001; 322 : 822 (doi :10.1136/bmj.322.7290.822).

FÖH, B., Borsche M, Balck A, Taube S, Rupp J, Klein C, Katalinic A. (2021). Complications of nasal and pharyngeal swabs: a relevant challenge of the COVID-19 pandemic? *European Respiratory Journal, 2021 Apr 8;57(4):*2004004. (doi : 10.1183/13993003.04004-2020. PMID : 33303542; PMCID : PMC7736753). (Consulté le 29 novembre 2022).

GABIS, Lidia V., Odelia Leon Attia, Mia Goldman, Noy Barak, Paula Tefera, GLIGORIĆ, Vukašin, Margarida Moreira da Silva, Selin Eker, Nieke van Hoek, Ella Nieuwenhuijzen, Uljana Popova, Golnar Zeighami (2021). The usual suspects: How psychological motives and thinking styles predict the endorsement of well-known and COVID-19 conspiracy beliefs, *Applied Cognitive Psychology*, Volume 35, Issue 5, May 2021, p. 1171-1181 (https://doi.org/10.1002/acp.3844).

GAILLE Marie, Philippe Terral, Philippe Askenazy, Regis Aubry, Henri Bergeron, et al. Les sciences humaines et sociales face à la première vague de la pandémie de Covid-19 -Enjeux et formes de la recherche. [Rapport de recherche] Centre National de la Recherche Scientifique; Université Toulouse III - Paul Sabatier. 2020. (révision arrêtée au 15 mars 2021). ⟨halshs-03036192v2⟩. (Consulté le 11 avril 2024).

GÓMEZ, Á., López-Rodríguez, L., Sheikh, H., Ginges, J., Wilson, L., Waziri, H., Vázquez, A., Davis, R., & Atran, S. (2017). The devoted actor's will to fight and the spiritual dimension of human conflict. *Nature human behaviour*, *1*(9), 673–679. (https://doi.org/10.1038/s41562-017-0193-3).

GOREIS A, Voracek M, "A systematic review and meta-analysis of psychological research on conspiracy beliefs: field characteristics, measurement instruments, and associations with personality traits", *Frontiers in Psychology*, 2019, 10 :205. (doi: 10.3389/fpsyg.2019.00205).

GREENBURGH Anna et Nichola J. Raihani (2022). Paranoia and conspiracy thinking. *Current Opinion in Psychology*, Vol 47. (https://doi.org/10.1016/j.copsyc.2022.101362)

GRUDNIEWICZ, Agnes et al. (2019). Predatory journals: no definition, no defence, *Nature, 576,* 2019, pp. 210-212 (*https://doi.org/10.1038/d41586-019-03759-y*) (Consulté le 23 septembre 2022).

GRUZD, A., & Mai, P. (2020). Going viral: How a single tweet spawned a COVID-19 conspiracy theory on Twitter. Big Data & Society, 7(2). (https://doi.org/10.1177/2053951720938405)

HAMID, N., Pretus, C., Atran, S., Crockett, M. J., Ginges, J., Sheikh, H., Tobeña, A., Carmona, S., Gómez, A., Davis, R., & Vilarroya, O. (2019). Neuroimaging 'will to fight' for sacred values: an empirical case study with supporters of an Al Qaeda associate. *Royal Society open science*, 6(6), 181585. (https://doi.org/10.1098/rsos.181585).

HAMID, Nafees (2021). The Neuroscience of 'Devoted Actors' Within Extremist Groups, *New Lines Magazine*, March 2021. (https://newlinesmag.com/writers/nafees-hamid/). (Pages consultées le 14 mai 2024).

HARDING, Karen L., Richard D., Judah, PhD; and Charles E. Gant (2003). Outcome-Based Comparison of Ritalin® versus Food-Supplement Treated Children with AD/HD, *Alternative Medicine Review, Vol. 8(3),* pp. 319-330.

HAUGAN, G., & Dezutter, J. (2021). Meaning-in-Life: A Vital Salutogenic Resource for Health. In G. Haugan (Eds.) et. al., *Health Promotion in Health Care – Vital Theories and Research.* (pp. 85–101). Springer. (PMID: 36315724), (Consulté le 10 juin 2024).

HUGHES S, Machan L. (2021). It's a conspiracy : Covid-19 conspiracies link to psychopathy, Machiavellianism and collective narcissism. *Personality and Individual Differences*, 171. (https://doi.org/10.1016/j.paid.2020.110559).

IMHOFF, Roland, Pia Karoline Lamberty (2017). Too special to be duped: Need for uniqueness motivates conspiracy beliefs, *European Journal of Social Psychology*, Vol. 47, Issue 6, Pages 724-734. (https://doi.org/10.1002/ejsp.2265).

IMHOFF, Roland, Pia Lamberty (2018). How paranoid are conspiracy believers? Toward a more fine-grained understanding of the connect and disconnect between paranoia and belief in conspiracy theories. *Special Issue: Belief in Conspiracy Theories as a Social-Psychological Phenomenon,* Vol. 48 (7), pp. 909-926. (https://doi.org/10.1002/ejsp.2494)

INSERM (2022). *Un récepteur synaptique impliqué dans l'émergence de croyances aberrantes*

INZLICHT, Michael, Alexa M. Tullett & Marie Good (2011): The need to believe: a neuroscience account of religion as a motivated process, *Religion, Brain & Behavior*, 1:3, 192-212

INZLICHT, Michael Alexa M. Tullett & Marie Good (2011): The need to believe: a neuroscience account of religion as a motivated process, *Religion, Brain & Behavior*, 1:3, 192-212.

JAKUB ŠROL, Eva, Ballová Mikušková, Vladimíra Čavojová (2021). When we are worried, what are we thinking? Anxiety, lack of control, and conspiracy beliefs amidst the COVID-19 pandemic,

Applied Cognitive Psychology, vol. 35, issue 3, pp. 720-729. (https://doi.org/10.1002/acp.3798).

JAMES, H., Marsh, (2021). L'épidémie de variole de Montréal en 1885. Dans : https://www.thecanadianencyclopedia.ca/fr/article/un-fleau-frappe-montreal. (Consulté le 7 juillet 2022).

JASINSKAJA-LAHTI, Inga, Jolanda Jetten (2019). Unpacking the relationship between religiosity and conspiracy beliefs in Australia, *Social Psychology,* Vol. 58 (4), October 2019, p. 938-954 (https://doi.org/10.1111/bjso.12314).

JOLLEY, D., & Douglas, K. M. (2014). The social consequences of conspiracism: Exposure to conspiracy theories decreases intentions to engage in politics and to reduce one's carbon footprint. *British Journal of Psychology, 105*(1), 35-56.

JONES, Richard Gardner (1969). A factored measure of Ellis's irrational belief system, with personality and maladjustment correlates. *Dissertation Abstracts International, 29*(11-B), 4379–4380.

JOYAL, Marilyne et Pascale Tremblay, *Le caractère essentiel de la réplication en sciences,* mars 2022, https://speechneurolab.ca/fr/blogue/item/312-le-caractere-essentiel-de-la-replication-en-sciences (Consulté le 16 septembre 2022).

KANOZIA, Rubal & Ritu Arya (2021) "Fake news", religion, and COVID-19 vaccine hesitancy in India, Pakistan, and Bangladesh, *Media Asia,* *48:4,* pp. 313-321, (DOI: 10.1080/01296612.2021.1921963)

KAUFLING, J., Freund-Mercier, M. J., & Barrot, M. (2016). Impact des opiacés sur les neurones dopaminergiques [Impact of opiates on dopaminergic neurons]. *Medecine sciences : M/S, 32*(6-7), 619–624. (https://doi.org/10.1051/medsci/20163206026)

KJÆRVIK, S. L., & Bushman, B. J. (2021). The link between narcissism and aggression: A meta-analytic review. *Psychological Bulletin, 147*(5), pp. 477-503. (https://doi.org/10.1037/bul0000323)

KOOB, G. F., & Le Moal, M. (2008). Review. Neurobiological mechanisms for opponent motivational processes in addiction. *Philosophical transactions of the Royal Society of London. Series B, Biological sciences, 363*(1507), 3113–3123. (https://doi.org/10.1098/rstb.2008.0094).

KRUGLANSKI, A. W. (2017). Why do People Believe Fake News? *HuffPost,* Sep 15, 2017.

KRUGLANSKI, A. W., & Fishman, S. (2009). The need for cognitive closure. In M. R. Leary & R. H. Hoyle (Eds.), *Handbook of individual differences in social behavior* (pp. 343–353). The Guilford Press. (https://psycnet.apa.org/record/2009-12071-023).

LANGER, E. J. (1975). The illusion of control. *Journal of Personality and Social Psychology, 32*(2), 311–328. (https://doi.org/10.1037/0022-3514.32.2.311)

LANTIAN, Anthony (2015). Rôle fonctionnel de l'adhésion aux théories du complot : un moyen de distinction ? *Thèse de doctorat en Sciences cognitives, psychologie et neurocognition*, Université Grenoble Alpes. (https://www.theses.fr/2015GREAS006).

LANTIAN, Anthony, Mike Wood, Biljana Gjoneska (2020). Personality traits, cognitive styles and worldviews associated with beliefs in conspiracy theories. Dans : *Butter M, Knight P (éditeurs.), Routledge Handbook of conspiracy theories, Routledge*, p. 157-167. (https://doi.org/10.4324/9780429452734).

LEE, J. Rajah, J. CSF (2021). Rhinorrhoea post COVID-19 swab: A case report and review of literature. CASE REPORT| Journal of Clinical Neuroscience, V. 86, P 6-9, APRIL 01, 2021).

LEIBOVITZ, Talia, Amanda L. Shamblaw, Rachel Rumas, Michael W. Best (2021). COVID-19 conspiracy beliefs: Relations with anxiety, quality of life, and schemas, *Personality and Individual Differences*, 175, June 2021. (https://doi.org/10.1016/j.paid.2021.110704).

LEVINE, Timothy R. (2022). Truth-default theory and the psychology of lying and deception detection

LUO, M., Hancock, J. T., & Markowitz, D. M. (2022). Credibility Perceptions and Detection Accuracy of Fake News Headlines on

Social Media: Effects of Truth-Bias and Endorsement Cues. *Communication Research,* *49*(2), 171–195. (https://doi.org/10.1177/00936502209211321).

LYONS-WEILER, James & Paul Thomas (2020). Relative Incidence of Office Visits and Cumulative Rates of Billed Diagnoses Along the Axis of Vaccination, *International Journal of Environmental Research and Public Health, 2020, 17(22)* : 8674. (doi: 10.3390/ijerph17228674).

MEHRA, Mandeep R., Sapan S. Desai, Frank Ruschitzka, Amit N. Patel (2020). RETRACTED: Hydroxychloroquine or chloroquine with or without a macrolide for treatment of COVID-19: a multinational registry analysis, *The Lancet* (https://doi.org/10.1016/S0140-6736(20)31180-6) (Consulté le 2 novembre 2022).

MENANT, Franck (2022). Les croyances à propos des comètes. *Futura Sciences,* 11-08-2022. (https://www.futura-sciences.com/sciences/dossiers/astronomie-cometes-ces-visiteuses-soir-138/page/4/) (Consulté le 15 août 2022).

ORANSKY, Ivan (2022). Retractions are increasing, but not enough, dans *Nature,* Août 2022, 608 (7921).

OVADIA, Daniela (2016) Festinger et la psychologie de l'incohérence (2016). Dans : *Cerveau & Psycho* N° 83, 16 novembre 2016. (https://www.cerveauetpsycho.fr/sd/psychologie/festinger-et-la-psychologie-de-l-incoherence-9366.php) (Consulté le 20 juillet 2022).

PALUCK E. L. (2009). Reducing intergroup prejudice and conflict using the media: a field experiment in Rwanda. *Journal of personality and social psychology*, *96*(3), 574–587. (https://doi.org/10.1037/a0011989).

PALUCK E. L. (2010). Is it better not to talk? Group polarization, extended contact, and perspective taking in Eastern Democratic Republic of Congo. *Personality & social psychology bulletin*, *36*(9), 1170–1185. (doi : 10.1177/0146167210379868).

PATTERSON, Grace E. K., Marie McIntyre, Helen E. Clough and Jonathan Rushton (2021). Societal Impacts of Pandemics: Comparing COVID-19 With History to Focus Our Response. *Frontiers Public Health*, Volume 9, Article 630449, 12 April 2021 (https://doi.org/10.3389/fpubh.2021.630449).

PENNYCOOK, G., Rand DG. (2121). The Psychology of Fake News. *Trends in Cognitive Science*, 25(5), p. 388-402. (doi: 10.1016/j.tics.2021.02.007).

PILCH I, Turska-Kawa A, Wardawy P, Olszanecka-Marmola A, Smołkowska-Jędo W. Contemporary trends in psychological research on conspiracy beliefs. A systematic review. Front Psychol. 2023 Feb 8; 14:1075779. doi: 10.3389/fpsyg.2023.1075779. PMID: 36844318; PMCID: PMC9945548.

POUPART, F.L., M.J.S. Bouscail (2021). Enjeux psychiques et psychopathologiques des croyances conspirationnistes : de la crise sanitaire du COVID-19 à la crise existentielle, *Annales Médico-psychologiques, revue psychiatrique,* Vol. 179 (4), April 2021, Pages 311-316. (https://doi.org/10.1016/j.amp.2021.03.004).

PRETUS, C., Hamid, N., Sheikh, H., Ginges, J., Tobeña, A., Davis, R., Vilarroya, O., & Atran, S. (2018). Neural and Behavioral Correlates of Sacred Values and Vulnerability to Violent Extremism. *Frontiers in psychology,* 9, 2462. https://doi.org/10.3389/fpsyg.2018.02462

RALUCA, Enescu (2011). Influence des Moyens de Preuve et de la Certitude du Verdict sur le Jugement Pénal, *Revue Internationale de criminologie et de police technique et scientifique, Vol. 3(11),* p. 267 à 280.

RIMÉ, B. (2019). Emotions at the Service of Cultural Construction, *Sage Journal, Vol. 12 (2),* p. 65-78. (https://doi.org/10.1177/1754073919876036)

RINALDI, R. (2020). Pourquoi préférons-nous les infox ? *Sciences Humaines,* 323, 4-4. (https://doi.org/10.3917/sh.323.0004) (Consulté le 2 novembre 2022).

ROSSMO, Kim (2020). Anatomie d'une enquête criminelle. *Criminologie,* Volume 53, numéro 2, p. 17-42. (https://doi.org/10.7202/1074187ar)

ROYER, Isabelle, DMSP Dauphine, « Escalade de l'engagement : décideurs et responsabilité : étude du cas "Les Amants du Pont Neuf" » https://www.strategie-aims.com/conferences/16/themes#communication_1005 (Consulté le 11 octobre 2022).

SACHS, Jeffrey D, Salim S Abdool Karim, Lara Aknin, Joseph Allen, Kirsten Brosbøl, Francesca Colombo, Gabriela Cuevas Barron, María Fernanda Espinosa, Vitor Gaspar, Alejandro Gaviria, Andy Haines, Peter J Hotez, Phoebe Koundouri, Felipe Larraín Bascuñán et al. (2022). The *Lancet* Commission on lessons for the future from the COVID-19 pandemic, *The Lancet, Vol 400 (10359),* p.1224-1280. (DOI : https://doi.org/10.1016/S0140-6736(22)01585-9).

SALVADOR, A., Arnal, L.H., Vinckier, F. *et al.* Premature commitment to uncertain decisions during human NMDA receptor hypofunction. *Nature Communications, 13,* 338 (2022). (https://doi.org/10.1038/s41467-021-27876-3) (Consulté le 6 janvier 2023).

SAROGLOU, V. (2010). Religiousness as a cultural adaptation of basic traits : a five-factor model perspective. *Pers Soc Psychol Rev.* 2010 Feb ; 14(1) :108-25. (doi : 10.1177/1088868309352322) (Consulté le 5 décembre 2020).

SAROGLOU, V. (2013). Prédisposé à croire ? *Le Monde de l'intelligence,* n°30, Avril-Mai 2013, p. 18-22.

(https://www2.psych.ubc.ca/~ara/Manuscripts/LMi30-dossier%20Dieu.pdf) (Consulté le 5 décembre 2022).

SHAHAR, Shefer, Meirav Shaham, Tally Lerman-Sagie, The myth of vaccination and autism spectrum, *European Journal of Paediatric Neurology,* *Vol.* *36,* Janvier 2022, P. 151-158 (https://doi.org/10.1016/j.ejpn.2021.12.011).

SHEIKH, H., Ginges, J., & Atran, S. (2013). Sacred values in the Israeli-Palestinian conflict: resistance to social influence, temporal discounting, and exit strategies. *Annals of the New York Academy of Sciences, 1299*, 11–24. (https://doi.org/10.1111/nyas.12275).

SHEIKH, H., Ginges, J., Coman, A., & Atran, S. (2012). Religion, group threat and sacred values. *Judgment and Decision Making, 7*(2), 110–118. (https://doi.org/10.1017/S1930297500002965).

Sheikh, H., Gómez, Á., & Atran, S. (2016). Empirical evidence for the devoted actor model. *Current Anthropology, 57*(S13), S204-S209 (DOI:10.1086/686221).

SIDERIS, Felicia, Les fausses infos se répandent plus vite que le coronavirus" : sur Facebook, "l'infodémie" explose, Publié le 26 février 2020 à 14h51, sur : https://www.tf1info.fr/sante/les-fausses-infos-se-repandent-plus-vite-que-le-coronavirus-sur-facebook-l-infodemie-explose-2146381.html. (Consulté le 16 juin 2022).

SIMIAND, François, " Méthode historique et science sociale. Étude critique d'après les ouvrages récents de M. Lacombe et de M.

Seignobos " (1903). Extrait de la Revue de synthèse historique, 1903, pp. 1-22. Texte reproduit dans l'ouvrage de François Simiand, Méthode historique et sciences sociales. (pp 113 à 137) Réimpression. Paris: Éditions des archives contemporaines, 1987, 534 pp. Choix de Marina Cedronio. (http://classiques.uqac.ca/classiques/simiand_francois/methode/methode_11/methode_hist_sc_soc1.html) (Consulté le 11 avril 2024).

SKINNER, B. F. (1948). "Superstition" in the pigeon. *Journal of Experimental Psychology,* Vol 38, 168-272. (https://doi.org/10.1037/h0055873).

SKODOL, Andrew (2020). Présentation des troubles de la personnalité, dans https://www.merckmanuals.com/fr-ca/accueil/troubles-mentaux/troubles-de-la-personnalité/présentation-des-troubles-de-la-personnalité. (Consulté le 25 décembre 2021).

SOLOMON, R.L. (1980). The opponent-process theory of acquired motivation: the costs of pleasure and the benefits of pain. *The American psychologist, 35(8)*, p. 691-712.

SOLOMON, Richard L. and John D. Corbit (1974). An opponent-process theory of motivation: I. Temporal dynamics of affect. *Psychological Review, 81(2)*, p. 119-145. (https://doi.org/10.1037/h0036128)

STANLEY, M.L., Whitehead, P.S., Marsh, E.J. *et al.* (2022). Prior exposure increases judged truth even during periods of mind wandering. *Psychonomic Bulletin & Review,* **29**, pp. 1997–2007. (https://doi.org/10.3758/s13423-022-02101-4)

STAW, Barry M. « Knee-deep in the big muddy : a study of escalating commitment to a chosen course of action », *Organizational Behavior and Human Performance*, vol. 16, n° 1, 1976, p. 27–44. (https://doi.org/10.1016/0030-5073(76)90005-2).

SULLIVAN, C. B. et al. (2019). Cerebrospinal Fluid Leak After Nasal Swab Testing for Coronavirus Disease 2019. *JAMA Otolaryngol Head Neck Surg. 2020 dec 1 ; 146(12) :* 1179-1181. 3.

SUSSMAN, S., Lisha, N., & Griffiths, M. (2011). Prevalence of the addictions: a problem of the majority or the minority?. *Evaluation & the health professions, 34(1),* 3–56. (https://doi.org/10.1177/0163278710380124)

SUSSMAN, S., Reynaud, M., Aubin, H. J., & Leventhal, A. M. (2011). Drug addiction, love, and the higher power. *Evaluation & the health professions, 34(3),* 362–370. (https://doi.org/10.1177/0163278711401002)

SYLOS-LABINI, Mauro et Natalia Zinovyeva, A walk on the wild side : 'Predatory' journals and information asymmetries in scientific evaluations, *Research Policy, Vol. 48(2),* 2019, Pages 462-477 (https://doi.org/10.1016/j.respol.2018.04.013).

TAYLOR, C.Z. (2002). Religious Addiction: Obsession with Spirituality. *Pastoral Psychology* **50**, 291–315 (2002). (https://doi.org/10.1023/A:1014074130084).

TAYLOR, Luke, E., Amy L Swerdfeger et Guy D Eslick, « Vaccines are not associated with autism : an evidence-based meta-analysis of case-control and cohort studies », *Vaccine*, Elsevier, vol. 32, nº 29, 17 juin 2014, p. 3623-9. (https://doi.org/10.1016/j.vaccine.2014.04.085).

THE EDITORS OF THE LANCET (2010). Retraction--Ileal-lymphoid-nodular hyperplasia, non-specific colitis, and pervasive developmental disorder in children », *Lancet*, vol. 375, nº 9713, février 2010, p. 445 (https://doi.org/10.1016/S0140-6736(10)60175-4).

TOSSEL, C. C., Gómez, A., de Visser, E. J., Vázquez, A., Donadio, B. T., Metcalfe, A., Rogan, C., Davis, R., & Atran, S. (2022). Spiritual over physical formidability determines willingness to fight and sacrifice through loyalty in cross-cultural populations. *Proceedings of the National Academy of Sciences of the United States of America, 119*(6), e2113076119. (https://doi.org/10.1073/pnas.2113076119).

TREMBLAY, André, Suicide, migration et rapports sociaux de sexe, *Recherches sociographiques, Vol. 48, No. 3,* septembre–décembre 2007, p. 65–96. (https://doi.org/10.7202/018004ar) (Consulté le 13 mars 2023).

VAN PROOIJEN, J.-W. (2021). How conspiracy theories bypass people's rationality, *Psyche*, 20 oct. 2021. (https://psyche.co/ideas/how-conspiracy-theories-bypass-peoples-rationality?fbclid=IwAR24IPEij-MqPHfmqHtUESVlLoNBMWz7ZInhB9CxThdKJjR65z6QXz83N9A). (Consulté le 16 décembre 2021).

VAN PROOIJEN, J.-W., Klein, O., & Milošević Đorđević, J. (2020). Social-cognitive processes underlying belief in conspiracy theories. Dans M. Butter & P. Knight (Eds.), *Handbook of Conspiracy Theories* (pp. 168-180). Oxon, Royaume-Uni : Routledge.

VAN PROOIJEN, Jan-Willem, K. M. Douglas and C. De Inocencio (2018). Connecting the dots : Illusory pattern perception predicts belief in conspiracies and the supernatural. *European Journal of Social Psychology*, Vol. 48, Issue 3, pp 320-335. (https://doi.org/10.1002/ejsp.2331).

VAN PROOIJEN, Jan-Willem, Karen M. Douglas (2018). Belief in conspiracy theories : Basic principles of an emerging research domain, *European Journal of Social Psychology, 8(7),* p. 897-908. (https://doi.org/10.1002/ejsp.2530)

VIALLON, P., Dolbeau-Bandin, C. & Picot, J. (2021). Introduction : L'infodémie entre information et désinformation. *Les Cahiers du numérique*, 17, 9-15. (https://doi.org/10.3166/LCN.2021.010)

VINCENDON, Salomé (2022) MULTIPLICATION DES TESTS NASOPHARYNGÉS : Y A-T-IL UN RISQUE À SE FAIRE BEAUCOUP TESTER ? Dans BFMTV, 02/02/2022 (https://www.bfmtv.com/sante/multiplication-des-tests-nasopharynges-y-a-t-il-un-risque-a-se-faire-beaucoup-tester_AN-202202020346.html) (Consulté le 22 novembre 2022).

VIREN, Swami, Martin Voracek, Stefan Stieger, Ulrich S. Tran, Adrian Furnham (2014). Analytic thinking reduces belief in conspiracy theories, *Cognition, Volume 133*, Issue 3, Pages 572-585. (https://doi.org/10.1016/j.cognition.2014.08.006).

VÎSLĂ A, Flückiger C, grosse Holtforth M, David D. Irrational Beliefs and Psychological Distress: A Meta-Analysis. *Psychother Psychosom.* 2016;85(1):8-15. (doi: 10.1159/000441231).

VÎSLĂ Andree, Christoph Flückiger, Martin grosse Holtforth, Daniel David (2016). Irrational Beliefs and Psychological Distress: A Meta-Analysis. *Psychother Psychosom.* 85(1):8-15. (doi: 10.1159/000441231).

VITAUD, Laetitia (2019). « Comment le biais des « coûts irrécupérables » vous rend irrationnel », dans Welcome to the Jungle, (https://www.welcometothejungle.com/fr/articles/biais-couts-irrecuperables), (Consulté le 24 février 2022).

VOSOUCHI, SOROUSE, Deb Roy and Sinan Aral (2018), The spread of true and false news online, *SCIENCE,* Vol 359, Issue 6380, pp. 1146-1151. (DOI : 10.1126/science.aap955).

WAGNER-EGGER, P. « Pourquoi les êtres humains croient-ils autant aux théories du complot ? 2) les causes psychologiques. », *Blick*, 10 novembre 2021. (https://www.blick.ch/fr/news/opinion/la-chronique-de-pascal-wagner-egger-pourquoi-les-etres-humains-croient-ils-autant-aux-theories-du-complot-2-3-id16976785.html).

WAGNER-EGGER, Pascal, dans https://www.cnews.fr/monde/2021-09-13/pascal-wagner-egger-le-complotisme-est-un-discours-de-revanche-contre-les-elites (Consulté le 7 janvier 2022).

WAGNER-EGGER, Pascal, *Psychologie des croyances aux théories du complot – les bruits de la conspiration*, Presses universitaires de Grenoble, 2021, p. 89.

WAGNER-EGGER, Pascal, *Quand les prophéties complotistes échouent*, dans Blick, 18 mai 2022. https://www.blick.ch/fr/news/opinion/pascal-wagner-egger-quand-les-propheties-complotistes-echouent-id17415265.html

WALBURG, Vera Solène Arnault et Stacey Callahan (2014). Les croyances rationnelles et irrationnelles en lien avec le niveau de stress des étudiants confrontés à l'idée d'un échec aux examens universitaires, *Journal de Thérapie Comportementale et Cognitive, Volume 24, Issue 1,* March 2014, Pages 14-23 (doi.org/10.1016/j.jtcc.2013.09.004)

WARD, C., & Voas, D. (2011). The Emergence of Conspirituality. *Journal of Contemporary Religion*, 26, pp. 103-121. (https://doi.org/10.1080/13537903.2011.539846).

WHISTON, J.A. and A.D. Galinsky (2008). Lacking Control Increases Illusory Pattern Perception, *SCIENCE*, Vol 322, Issue 5898, pp. 115-117. (https://doi.org/10.1126/science.1159845).

WILMES, A. (2014). Le concept de psychopathie est-il cohérent ? Bases cérébrales et responsabilité morale. *PSN*, 12, 31-49. (https://doi.org/10.3917/psn.121.0031).

WOOD, M.J., Douglas, K. M., Sutton, R. M. (2012) Dead and alive: beliefs in contradictory conspiracy theories. *Social Psychological and Personality Science 3*: 767–773.

YOSHIJA Walter, and Andreas Altorfer (2023). Electrodermal Activity Implicating a Sympathetic Nervous System Response under the Perception of Sensing a Divine Presence, A Psychophysiological Analysis. *Psych* 2023, *5*(1), 102-112. (https://doi.org/10.3390/psych5010010).

YOUNG, K. S., van der Velden, A. M., Craske, M. G., Pallesen, K. J., Fjorback, L., Roepstorff, A., & Parsons, C. E. (2018). The impact of mindfulness-based interventions on brain activity: A systematic review of functional magnetic resonance imaging studies. *Neuroscience and biobehavioral reviews, 84*, 424–433. (https://doi.org/10.1016/j.neubiorev.2017.08.003).

YUSTISIA, W., Putra, I. E., Kavanagh, C., Whitehouse, H., & Rufaedah, A. (2020). The role of religious fundamentalism and tightness-looseness in promoting collective narcissism and extreme group behavior. *Psychology of Religion and Spirituality,* Vol. 12(2), pp. 231-240. (https://doi.org/10.1037/rel0000269)

ZAKARY L., Tormala, Derek D. Rucker. Attitude certainty : Antecedents, consequences, and new directions. *Consumer Psychology Review, Vol. 1(1),* 2017, pp. 72-89. (https://doi.org/10.1002/arcp.1004).

ZHOU, Y., & Shen, L. (2022). Confirmation Bias and the Persistence of Misinformation on Climate Change. *Communication Research*, *49*(4), 500–523. (https://doi.org/10.1177/00936502211028049)

ZIMMERMAN, Mark. Trouble de la personnalité paranoïde. Dans https://www.merckmanuals.com/fr-ca/professional/troubles-psychiatriques/%EF%BB%BFtroubles-de-la-personnalité/trouble-de-la-personnalité-paranoïde?query=paranoïaque. (Consulté le 30 juin 2022).

ZIMMERMAN, Mark. Troubles de la personnalité dépendante, dans https://www.merckmanuals.com/fr-ca/professional/troubles-psychiatriques/%EF%BB%BFtroubles-de-la-personnalité/trouble-de-la-personnalité-dépendante. (Consulté le 21 juin 2022).

SITES INTERNET

http://1libertaire.free.fr/Soumission08.html (Consulté le 11 octobre 2022).

http://www.psychomedia.qc.ca/lexique/definition/heuristique (Consulté le 15 octobre 2022).

https://7zine.com/lintuition-nous-laisse-souvent-tomber-comment-utiliser-les-probabilites-et-les-statistiques-pour-trouver-les-vraies-reponses/ Consulté le 27 septembre 2022).

https://couleurs-de-la-vie.blog4ever.com/ecouvillons-danger-protegez-votre-barriere-hemato-encephalique (Consulté le 23 novembre 2022).

https://culturesciences.chimie.ens.fr/thematiques/chimie-organique/chimie-pharmaceutique/de-la-molecule-au-medicament. (Consulté le 10 septembre 2022).

https://dicophilo.fr/definition/inference/ (Consulté le 11 juillet 2022).

https://dpa-factchecking.com/belgium/200717-99-833567/ (Consulté le 22 novembre 2022).

https://en.wikipedia.org/wiki/InfoWars (Consulté le 7 novembre 2022).

https://factuel.afp.com/doc.afp.com.329C4KE (Consulté le 25 novembre 2022).

https://fr.statista.com/infographie/28536/evolution-esperance-de-vie-dans-le-monde-et-dans-une-selection-de-pays-g20/ (Consulté le 13 mars 2023).

https://fr.wikipedia.org/wiki/Bien_fongible. (Consulté le 30 avril 2024).

https://fr.wikipedia.org/wiki/Controverse_sur_le_rôle_de_la_vaccination_dans_l%27autisme (Consulté le 16 septembre 2022).

https://fr.wikipedia.org/wiki/Critique_de_la_philosophie_du_droit_de_Hegel (Consulté le 17 avril 2024).

https://fr.wikipedia.org/wiki/Désinformation_sur_la_pandémie_de_Covid-19. (Consulté le 7 juillet 2022).

https://fr.wikipedia.org/wiki/Dictature_sanitaire (Consulté le 25 novembre 2022).

https://fr.wikipedia.org/wiki/Effet_Stroop

https://fr.wikipedia.org/wiki/Empirisme (Consulté le 17 août 2022).

https://fr.wikipedia.org/wiki/Épidémie_de_variole_de_Montréal_en_1885. (Consulté le 5 juillet 2022).

https://fr.wikipedia.org/wiki/Esp%C3%A9rance_de_vie_humaine#Facteurs_%C3%A9conomiques_et_socio%C3%A9conomiques (Consulté le 13 mars 2023).

https://fr.wikipedia.org/wiki/Faux_négatif (Consulté le 23 août 2022).

https://fr.wikipedia.org/wiki/Faux_positif (Consulté le 23 août 2022).

https://fr.wikipedia.org/wiki/Flower_Power. (Consulté le 2 février 2023).

https://fr.wikipedia.org/wiki/Homologation (Consulté le 10 janvier 2023).

https://fr.wikipedia.org/wiki/Lashkar-e-Toiba (Consulté le 14 mai 2024).

https://fr.wikipedia.org/wiki/Liste_d%27avions_civils_abattus (Consulté le 10 janvier 2023).

https://fr.wikipedia.org/wiki/New_Age#Histoire. (Consulté le 2 février 2023).

https://fr.wikipedia.org/wiki/Offuscation#:~:text=L'offuscation%2C %20assombrissement%2C%20obscurcissement,peut%20%C3%AAt re%20intentionnelle%20ou%20involontaire. (Consulté le 11 avril 2024).

https://fr.wikipedia.org/wiki/Probabilité_conditionnelle (Consulté le 26 septembre 2022).

https://fr.wikipedia.org/wiki/Problème_de_la_démarcation (Consulté le 24 août 2022).

https://fr.wikipedia.org/wiki/Publier_ou_périr (Consulté le 10 janvier 2023).

https://fr.wikipedia.org/wiki/Retraction_Watch (Consulté le 10 janvier 2023).

https://fr.wikipedia.org/wiki/Test_statistique (Consulté le 14 septembre 2022).

https://gdt.oqlf.gouv.qc.ca/ficheOqlf.aspx?Id_Fiche=17032988 (Consulté le 22 octobre 2022).

https://geneticliteracyproject.org/glp-facts/russell-blaylock/ (Consulté le 7 novembre 2022).

https://ici.radio-canada.ca/nouvelle/1702335/depistage-covid-19-precision-foyer-centre-soin-longue-duree-ontario-nord (Consulté le 10 janvier 2023).

https://ici.radio-canada.ca/nouvelle/1830655/opposants-vaccination-comparaison-mesures-sanitaires-nazisme (Consulté le 20 novembre 2022).

https://ici.radio-canada.ca/nouvelle/1867818/nazi-ukraine-propagande-russe-azov-poutine (Consulté le 8 mai 2024).

https://ici.radio-canada.ca/nouvelle/1914864/pandemie-covid-coronavirus-gestion-lancet

https://lecerveau.mcgill.ca/flash/a/a_03/a_03_cl/a_03_cl_par/a_03_cl_par.html#:~:text=La%20stimulation%20du%20noyau%20accumbens,la%20quantit%C3%A9%20de%20neurotransmetteurs%20rel%C3%A2ch%C3%A9s.

https://publicationethics.org (Consulté le 20 septembre 2022).

https://pubpeer.com (Consulté le 19 septembre 2022).

https://pubpeer.com/publications/16786B986703B5373DFBCAF1D4BCCC. (Consulté le 7 novembre 2022).

https://pubpeer.com/publications/9A93419B03160444B7AFF65411D228 (Consulté le 7 novembre 2022).

A Rabbit in the Clouds

https://pubpeer.com/search?q=%22The+UK+accelerated+immunisation+programme+and+sudden+unexpected+death+in+infancy%3A+case-control+study%22 (Consulté le 30 septembre 2022).

https://pubpeer.com/search?q=doi%3A+10.3390%2Fijerph17228674 (Consulté le 30 septembre 2022).

https://qactus.fr/2021/12/29/canada-j-trudeau-le-psychopathe-fasciste-maxime-bernier/ (Consulté le 20 novembre 2022).

https://retractionwatch.com (Consulté le 19 septembre 2022).

https://retractionwatch.com/?s=The+UK+accelerated+immunisation+programme+and+sudden+unexpected+death+in+infancy%3A+case-control+study (Consulté le 30 septembre 2022).

https://retractionwatch.com/2021/08/11/authors-blame-a-ghoul-for-retraction-of-paper-claiming-vaccines-lead-to-health-and-behavioral-issues/ (Consulté le 30 septembre 2022).

https://scienceetonnante.com/2012/10/08/les-probabilites-conditionnelles-bayes-level-1/ (Consulté le 1er septembre 2022).

https://spharm-inc.com/fr/le-processus-dexamen-et-dapprobation-des-medicaments-au-canada-un-guide-numerique/ (Consulté le 19 octobre 2022).

https://theoutline.com/post/1183/the-quack-behind-the-msg-scare-is-still-stoking-fear-for-profit (Consulté le 7 novembre 2022).

https://thevaccinereaction.org/2022/01/mayo-clinic-fires-700-employees-for-refusing-to-get-covid-19-vaccinations/ (Consulté le 7 novembre 2022).

https://twitter.com/cherielle100/status/1360389245635067905 (Consulté le 20 novembre 2022).

https://twitter.com/maximebernier/status/1431605652389081092?lang=fr

https://unpeudairfrais.org/11-vaccins-appel-urgent-aux-personnels-et-professionnels-de-sante/ (Consulté le 30 septembre 2022).

https://unpeudairfrais.org/relative-incidence-of-office-visits-and-cumulative-rates-of-billed-diagnoses-along-the-axis-of-vaccination/ (Consulté le 30 septembre 2022).

https://web.archive.org/web/20120630205839/http://voices.washingtonpost.com/checkup/2011/01/wakefield_tried_to_capitalize.html (Consulté le 20 octobre 2022).

https://www.11vaccinsobligatoires.com/appel-montagnier-joyeux/script/ (Consulté le 30 septembre 2022).

https://www.academie-medecine.fr/les-prelevements-nasopharynges-ne-sont-pas-sans-risque/ (Consulté le 22 novembre 2022).

https://www.canada.ca/fr/sante-canada/services/science-recherche/avis-scientifiques-processus-decisionnel/comite-ethique-recherche/a-propos-nous.html (Consulté le 19 octobre 2022).

A Rabbit in the Clouds

https://www.canada.ca/fr/sante-canada/services/securite-produits-consommation/pesticides-lutte-antiparasitaire/titulaires-demandeurs/homologation-nouveaux-produits.html (Consulté le 1er mars 2023).

https://www.chop.edu/centers-programs/vaccine-education-center/vaccines-and-other-conditions/vaccines-sudden-infant-death-syndrome-sids (Consulté le 30 septembre 2022).

https://www.cnrtl.fr/definition/anti-intellectualisme (Consulté le 13 juin 2022).

https://www.crrf-fcrr.ca/fr/bibliotheque/glossaire-fr-fr-1/item/27179-croyance-systemique (Consulté le 2 mars 2023).

https://www.dw.com/en/how-denial-and-conspiracy-theories-fuel-coronavirus-crisis-in-pakistan/a-53913842 (Consulté le 17 octobre 2022).

https://www.fda.gov/ (Consulté le 1er mars 2023).

https://www.hammadsheikh.org/research.html (Consulté le 30 avril 2024).

https://www.happyneuronactiv.com/articles/fonctions-cognitives-definition/#:~:text=La%20stimulation%20cognitive%20sollicite%20certains,allons%20pr%C3%A9senter%20de%20mani%C3%A8re%20succincte. (Consulté le 14 mai 2024).

https://www.huffpost.com/entry/why-do-people-believe-fake-news_b_59bc73dde4b06b71800c396f (Consulté le 4 octobre 2022).

https://www.inserm.fr/dossier/medicament-developpement/ (Consulté le 13 septembre 2022).

https://www.jmp.com/fr_ca/statistics-knowledge-portal/what-is-correlation.html (Consulté le 3 novembre 2022).

https://www.lapresse.ca/international/asie-et-oceanie/2023-12-20/coree-du-nord/kim-jong-un-promet-une-riposte-atomique-si-provoquee-par-des-armes-nucleaires.php (Consulté le 8 mai 2024).

https://www.larousse.fr/dictionnaires/francais/anecdote/3396 (Consulté le 21 novembre 2022).

https://www.le-cortex.com/article/les-heuristiques-ces-raccourcis-de-pensee-quotidiens. (Consulté le 14 juillet 2022).

https://www.ledevoir.com/monde/moyen-orient/805703/netanyahou-reaffirme-refus-souverainete-palestinienne-gaza (Consulté le 8 mai 2024).

https://www.leem.org/recherche-et-developpement (Consulté le 1er novembre 2022).

https://www.legifrance.gouv.fr/codes/section_lc/LEGITEXT000006072665/LEGISCTA000006155273/#:~:text=L'homologation%20ne%20peut%20%C3%AAtre,la%20qualit%C3%A9%20de%20la%20fabrication. (Consulté le 1er mars 2023).

https://www.lesechos.fr/idees-debats/editos-analyses/coronavirus-anatomie-dune-hysterie-collective-1181776 (Consulté le 20 novembre 2022).

https://www.liberation.fr/france/2017/11/07/du-mauvais-theatre-contre-les-vaccins_1608462/ (Consulté le 29 septembre 2022).

https://www.mdpi.com/1660-4601/18/15/7754 (Consulté le 30 septembre 2022).

https://www.mdpi.com/1660-4601/18/3/936 (Consulté le 30 septembre 2022).

https://www.medecinesciences.org/en/articles/medsci/full_html/2005/03/medsci2005213p315/medsci2005213p315.html (Consulté le 27 septembre 2022).

https://www.medpagetoday.com/special-reports/exclusives/98940 (Consulté le 7 novembre 2022).

https://www.museedelhistoire.ca/blog/vaccination-obligatoire/ (Consulté le 30 novembre 2022).

https://www.numerama.com/sciences/614708-chloroquine-les-graves-erreurs-scientifiques-de-la-methode-raoult.html

https://www.nytimes.com/2016/10/08/us/donald-trump-tape-transcript.html (Consulté le 8 mai 2024).

https://www.occitanie.ars.sante.fr/11-vaccins-obligatoires-pour-les-0-2-ans-ce-qui-change-en-2018-2 (Consulté le 30 septembre 2022).

https://www.revmed.ch/revue-medicale-suisse/2019/revue-medicale-suisse-664/comment-rediger-un-case-report (Consulté le 29 novembre 2022).

https://www.stresshumain.ca

https://www.thecanadianencyclopedia.ca/fr/article/revolution-tranquille (Consulté le 11 janvier 2023).

https://www.thelancet.com/journals/lancet/article/PIIS0140-6736(04)15715-2/

https://www.uottawa.ca/gazette/fr/nouvelles/comprendre-combattre-effets-du-stress-cerveau-sante-mentale#:~:text=Du%20c%C3%B4t%C3%A9%20du%20cortex%20pr%C3%A9frontal,s'en%20trouve%20alors%20affect%C3%A9e. (Consulté le 1ᵉʳ février 2023).

235

Printed in the United States

Legal deposit: Fourth quarter 2026

Distribution and dissemination:

Global: KDP

Made in the USA
Monee, IL
07 July 2026

56544286R00134